Lecture Notes on Multidisciplinary Industrial Engineering

Series Editor

J. Paulo Davim, Department of Mechanical Engineering, University of Aveiro, Aveiro, Portugal

"Lecture Notes on Multidisciplinary Industrial Engineering" publishes special volumes of conferences, workshops and symposia in interdisciplinary topics of interest. Disciplines such as materials science, nanosciences, sustainability science, management sciences, computational sciences, mechanical engineering, industrial engineering, manufacturing, mechatronics, electrical engineering, environmental and civil engineering, chemical engineering, systems engineering and biomedical engineering are covered. Selected and peer-reviewed papers from events in these fields can be considered for publication in this series.

More information about this series at http://www.springer.com/series/15734

María Teresa Castañeda Galvis ·
Jairo Nuñez Rodriguez ·
María del Coral Pérez Ordoñez ·
Marcela Villa Marulanda
Editors

Proceedings of the International Congress of Industrial Engineering (ICIE2019)

Engineering Paradigm of Change

Springer

Editors
María Teresa Castañeda Galvis
Universidad Pontificia Bolivariana
Seccional Bucaramanga, Santander, Colombia

María del Coral Pérez Ordoñez
Universidad Pontificia Bolivariana
Seccional Bucaramanga, Santander, Colombia

Jairo Nuñez Rodriguez
Santander
Universidad Pontificia Bolivariana
Seccional Bucaramanga, Santander, Colombia

Marcela Villa Marulanda
Universidad Pontificia Bolivariana
Seccional Bucaramanga, Santander, Colombia

ISSN 2522-5022 ISSN 2522-5030 (electronic)
Lecture Notes on Multidisciplinary Industrial Engineering
ISBN 978-3-030-49369-1 ISBN 978-3-030-49370-7 (eBook)
https://doi.org/10.1007/978-3-030-49370-7

This Springer imprint is published by the registered company Springer Nature Switzerland AG
The registered company address is: Gewerbestrasse 11, 6330 Cham, Switzerland

Preface

We are pleased to preface this book, in the framework of the 25 years of industrial engineering faculty celebration from the Pontificia Bolivariana University, Bucaramanga sectional, the International Industrial Engineering Congress ICIE 2019 were developed, promoted by the investigation groups Prolog (Production and Logistics) and Geetic (Enterprise, Education and TIC).

The congress main topic was: "Logistics, supply chain sustainability," the selected documents are recent and relevant investigations about industrial engineering and it application on the enterprise context, also the improvement on the people quality life.

On this book, you can find theoretical and practical contributions organized in three themes:

- I. Logistics
- II. Supply chain
- III. Sustainability

We hope this book offer an opportunity to enjoy a multidisciplinary vision.

October 2019

Maria Teresa Castañeda
Jairo Núñez Rodriguez

Organization

Program Committee

María Teresa Castañeda Galvis	Universidad Pontificia Bolivariana, Colombia
Marcela Villa Marulanda	Universidad Pontificia Bolivariana, Colombia
Jairo Núñez Rodriguez	Universidad Pontificia Bolivariana, Colombia
Cristopher Mejia Argueta	MIT Center for Transportation and Logistics, USA
María del Coral Pérez	Universidad Pontificia Bolivariana, Colombia
Josué Velázquez Martínez	MIT Center for Transportation and Logistics, USA
Daniel Rodriguez Sanchez	MIT Center for Transportation and Logistics, USA
Ludym Jaimes Carrillo	Universidad Pontificia Bolivariana, Colombia

Scientific Committee

International

Josué Velázquez Martínez	Massachusetts Institute of Technology, USA
Mara Guadalupe Moyano	Universidad del Valle de Atemajac, México
Natacha Pino Acuña	Universidad de Aysén, Chile
Hanzel Grillo	Universidad Costa Rica, Costa Rica
Ximena Rojas Lema	Escuela Politécnica Nacional, Ecuador

Nationals

Andrés Acevedo Ojeda	Universidad Pontificia Bolivariana, Colombia
Marco Antonio Villamizar Araque	Universidad Pontificia Bolivariana, Colombia

Oscar Orlando Ortíz Rodríguez	Universidad de Pamplona, Colombia
Rafael Guillermo García Cáceres	Universidad Pedagógica y Tecnológica de Colombia
Maryory Patricia Villamizar León	Universidad Pontificia Bolivariana, Colombia
Alba Soraya Aguilar Jiménez	Universidad Pontificia Bolivariana, Colombia
Orlando Federico González Casallas	Universidad Pontificia Bolivariana, Colombia

Contents

Sustainability

Logistic - Supply Chain - Sustainability

Logistic

Proposal to Characterize a Reverse Logistics Network for the Tetra Pak Package in the Municipality of Armenia, Quindío

Jhully P. Martínez[1](✉), Kaliyan Mathiyazhagan[2], Jose A. Solanilla[3], and Jhon G. Morales[3]

[1] Facultad de Ingeniería, Institución Universitaria EAM., Armenia, Colombia
jpmartinez@eam.edu.co
[2] Amity University, Noida, India
madii1984@yahoo.com
[3] Institución Universitaria EAM., Armenia, Colombia
{jsolanilla,jmorales}@eam.edu.co

Abstract. In Colombia, more than 900 million Tetra Pak packages are consumed in every year. The area with the highest consumption is the coffee axis, where there is neither a system of large-scale use of this waste nor a culture of separation at the source by the generator. It was found that educational institutions are large generators for this waste. The main project goal is characterizing the Tetra Pak consumption and determine the logistics network actors for the municipality of Armenia. The results show that principal actors are the load generator, the producer, the marketer (wholesaler and retailer), the delivery points, collection and classification centers, as well as recovery companies and companies with potential for use. Through a bibliographic review of 61 articles, the barriers faced by the proposed stakeholders were detected: delay and variability in returns, losses in quality, lack of strategic planning, etc.

Keywords: Characterization · Reverse logistics · Barriers · Tetra Pak · Quindío · Colombia

1 Introduction

Tetra Pak is a waste with a high generation volume and a wide potential for exploitation. It has been successfully employed in food preservation due to its composition: cardboard (75%), low density polyethylene (20%) and aluminum (5%). Worldwide, 188,000 million Tetra Pak containers were produced by 2016 in more than 175 countries (Tetra Pak, 2017). In the Colombian case for the year 2016, 900 million containers were produced (Alimentos 2019). In the city of Armenia, there are no consumption rates for Tetra Pak. However, there is a solid waste generation rate of 214.25 Tons/day, of which about 70% are potentially usable. It is known that the highest percentage of Tetra Pak package is generated in homes and educational institutions and is not collected yet, generating a

M. T. Castañeda Galvis et al. (Eds.): ICIE 2019, LNMUINEN, pp. 3–10, 2021.
https://doi.org/10.1007/978-3-030-49370-7_1

loss to obtain Aluminum, pellets or corrugated paper, bricks, chipboard, energy recovery, among others.

Thus, to achieve an increase in the recovery of solid waste and its subsequent utilization is necessary to implement reverse logistics strategies. In other words, to achieve a perfectly managed reverse flow that collects and classifies them in order to re-dispose them at a point where the highest economic value and the least environmental impact are obtained (Ilgin and Gupta 2010). It is necessary to establish new management and use policies, so that the amount of solid waste that reaches final disposal is less, since the capacity of the final disposal sites is at the limit.

According to the Final Solid Waste Disposal Report, for the year 2017 in Colombia a total of 10,327,551 Tons of Solid Waste were generated, of which only 11.7% is recycled (Superintendencia de Servicios Públicos Domiciliarios and Departamento Nacional de Planeación 2017). These statistics are discouraging by considering that the expected growth is 20% in the generation of solid waste, for the next ten years. Concerning to landfills, which is the most common long-term provision, exist a bad outlook. 7.5% are on the edge of their capacity, situation of Bucaramanga, Armenia, and Manizales; 15% have less than three years left; 28.1% may last between 3 and 10 years, such as the Doña Juana landfill in Bogotá, and only 35.6% could remain for more than a decade (Noguera and Olivero-Verbel 2010).

Consequently, the objective of this research progress is to present a proposal for the characterization of the reverse logistics network for the Tetra Pak container generated in the municipality of Armenia. This paper is structure as follow goals: identify stakeholders in the chain of responsibility for Tetra Pak waste. Quantify and classify the Tetra Pak generated at a point of interest in the city and determine the impact degree on the pressures that stakeholders facing to implement a reverse logistics network for the Tetra Pak package.

2 Methodology

The proposed methodology has two key components in its development. The first component included the identification of stakeholders, as well as the barriers they face and the pressures that guide the implementation of a reverse logistics network for Tetra Pak and the second involved the characterization of the population generating this waste. To determine the barriers and guidelines that can guide the stakeholders of the Tetra Pak chain, a literature review was carried out with the keywords: reverse logistics, implementation, barriers, and drivers. The search was done through the Science Direct and Scopus databases, as well into the Google Scholar search engine, in Spanish and English.

The second component includes construction, validation, and launch of the survey for the target population. In this case, it is an educational institution, in order to determine the amount and type of Tetra Pak waste generated, as well as the viability of a collection and classification point. The proposed survey consists of 16 Likert-scale questions and an open question to know the incentives that the population requires to dispose of the Tetra Pak at a collection point. Finally, the components are joined to obtain the construction of the reverse logistics network for the Tetra Pak in the city of Armenia, based on the stakeholders identified in the literature review and the characteristics of the Tetra Pak waste generation population.

3 Results

(i) Stakeholders identification in a reverse Logistic Network for the Tetra Pak packaging

The review in the academic literature to determinate stakeholders for the Tetra Pak reverse logistic network was carried out in the Science Direct, Scopus databases, in government documents, postgraduate academic theses and finally in the Google Scholar search engine. The results filtered by keywords are:

Table 1. List of articles obtained

Keyword	Keyword	Results	Final selection
Reverse logistics	Tetra Pak	37	3
Reverse logistics	Barriers	131	15
Reverse logistics	Implementation	296	27
Reverse logistics	Drivers	144	16

Final selection in Table 1, are those papers that are directly related to the topic of interest and with open access. Once the articles are filtered, an identification is made based on the results in the academic literature and some national normatives, as the Decreto 4741 of 2005, which refers to those responsible for the proper management of hazardous waste. Despite the fact Tetra Pak is not a hazardous waste, it shares some actors in the reverse logistics chain such as producer, market, control entities, collectors, among others. All stakeholders, as well as the relationships among them and their role in the potential use of Tetra Pak, are represented in the Fig. 1.

The actors that integrates the reverse logistics network are similar to those of the traditional supply chain: suppliers, producers, distributors, and marketers, which are categorized in Fig. 1 as TL Actor (Traditional Logistics actor). Furthermore, stakeholders such as the market, customers and social institutions are included, in addition to recyclers, cooperatives and recovery companies (RL Actor). According to Olejnik and Werner-Lewandowska (2018), national government creates specific laws to pressure other stakeholders. Subsequently, there are the traditional Supply Chain Actors (supplier, manufacturer, wholesaler, retailer, end user) and specialized reverse chain actors, for example, recycling and collection specialists, who are responsible for the advanced recovery. Finally, there is the society that drives the recovery and generates pressure on the government to create new protective laws.

In the case of recovery companies, in Colombia Tetra Pak estimates about seven plants in operation with the capacity to handle 90% of the Tetra Pak waste, but they

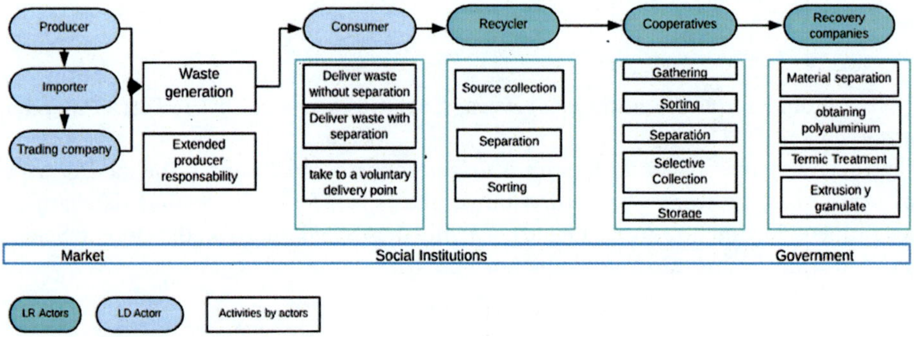

Fig. 1. Stakeholders in the reverse logistics network for Tetra Pak

have a maximum capacity of 280 Ton /month and recover 10% of the waste generated in Bogotá, Medellin, and Cali. Moreover, there are only 10 collection points for Bogotá and four for Cali. On the coffee axis and especially in Armenia, there are no plants for the collection, classification, and use of Tetra Pak, despite being an area with ample milk consumption.

(ii)Tetra Pak quantification and classification for an educational institution
The characterization survey was applied to 111 people belonging to an educational institution with roles as, teacher, student and administrative. Most of the study population live in families made up of between three and five persons. The most consumed Tetra Pak products are milk and juices, with an average per family of 24 and 5 units, respectively. Which generates an approximate consumption of 2160 containers. The consumption of drinks in Tetra Pak containers in the Educational Institution is mostly juices and chocolate drinks. Of these, the average consumption is two drinks per day per person. Currently, 66% of these containers are disposed incorrectly with plastic waste and 32% classifies it with ordinary waste since it contains drinks, only 2% throw it into the street. To avoid this inadequate provision, a collection point was established within the institution. Regarding waste management, 45% do not separate them from ordinary waste, but would be willing to separate them and 28% are separated at the source but are not collected. Among the incentives that the population expects to make the adequate disposition of the Tetra Pak are: the satisfaction of helping the environment (27%), accumulation of points in the gym and parking services (22%) and delivery of sweets (11%).

In the next phase of the investigation, the quantity of containers collected per month will be determined, at the same time that sensitization strategies are carried out against the use of Tetra Pak and the importance of disposing it in the selective collection container. Thus, it is evident that in the city of Armenia, there is a potential for the use of Tetra Pak and a willingness on the part of load generators, in this case, people linked to educational institutions.

(iii) Pressures facing stakeholders in the reverse logistics chain Two types of pressures were detected, internal and external. Internal pressures are specific to each link in the supply chain and generated by internal customers. Internal pressures includes restrictions on the supplier by the producer, the desire to reduce the risk in the processes by the

Table 2. Pressures generated by actors and stakeholders towards the supply chain

Actor	Pressures	Author
Supplier	Changes in the quality of the final product by including return material	(Rachih et al. 2019; Pinheiro et al. 2019)
	Shortage of technical information	(Shaharudin et al. 2015)
	Lack of material availability	(Govindan et al. 2015)
	Long-term benefits associated with the adoption of green strategies	(Mathiyazhagan and Haq 2013)
Producer	Effective management of reverse logistics operations	(Battaïa and Gupta 2015; Qiu et al. 2018)
	Higher profitability by reducing transport, inventory and storage costs	(Battaïa and Gupta 2015)
	Own resources and economic restrictions	(Shaharudin et al. 2015)
	Lack of senior management commitment	(Diabat and Govindan 2011a)
	Extended producer responsibility	(Diabat and Govindan 2011a; Shaharudin et al. 2015)
	Lack of integration of chain links	(Rubio Lacoba 2003)
	Reduction of environmental accidents in the organization	(Mathiyazhagan and Haq 2013)
Distributors	Occupancy of storage spaces and transport capacity	(Battaïa and Gupta 2015; Govindan and Popiuc 2014)
	High storage cost	(Shaharudin et al. 2015; Qiu et al. 2018)
	Lack of a collection routine	(Govindan et al. 2015; Pinheiro et al. 2019)
Costumer	Lack of incentives for customers	(Narayana et al. 2019)
	Lack of awareness about green logistics	(García Olivares 2004)
	Difficulty adopting green practices	(Shaharudin et al. 2015)
	Support and guide for return materials	(Olugu et al. 2011; Shaharudin et al. 2015)
	Demand from customers for environmental protection requirements	(Mathiyazhagan and Haq 2013)

(*continued*)

Table 2. (*continued*)

Actor	Pressures	Author
Market	Expiration and accumulation of disposable products in the market	(Narayana et al. 2019; Panigrahi et al. 2018)
	Accelerated market growth	(Diabat and Govindan, 2011a; Zhagan et al. 2019)
	Green Image	(Rubio Lacoba 2003)
Social Institutions	Commercial legislation that forces to take charge of the products	(Govindan et al. 2015; Govindan and Popiuc 2014)
	High costs at an adequate disposal	(García Olivares 2004; Panigrahi et al. 2018)
	Inadequate Government Policies	(Diabat and Govindan 2011a; Mathiyazhagan et al. 2019)
	Lack of funding to support RL activities	(Rachih et al. 2019; Pinheiro et al. 2019)
Society	Conduct codes	(Liao 2018)
	Lack of knowledge and awareness of environmental impacts	(Govindan et al. 2015)
	Lack of reverse logistics professionals	(Rubio Lacoba 2003)
	Greater increase in resource shortage	(Mathiyazhagan and Haq 2013)

worker, the committed vision of senior management, among others. External pressures are generated by stakeholders towards the links of the supply chain, such as the pressures of the end customer on the market through social institutions. The pressures of the competition, the pressures that society generates when analyzing the risks of improper handling of materials. The general list detected in the literature review is detailed in the following table.

According to Table 1, the pressures can be economic, operational, management, environmental and regulatory. In the case of the producer, the most obvious pressure is related with efficient management resources and the information system, as well as the lack of knowledge due to the economic impact generated by the return and changes in production systems. The society as an interested party, which generates pressures and limitations from the failures in the processes of behavior for the return of the material, given from the ignorance of the environmental impact, but at the same time it is aware of the scarcity of resources and demands more sustainable processes.

4 Discussion

Currently, there are successful management strategies for companies and supply chains such as the Triple Bottom line (Klewitz and Hansen 2014), Closed Supply chain (Liu and Chang, 2017), Green Supply Chain Management (Fahimnia et al. 2015), Green Manufacturing (Seth et al. 2018), implementation of environmental standards (Saengsupavanich et al. 2009), stewardship product, cleaner production (Khalili et al. 2015), among others. However, these methods and methodologies propose minimization from the generation or management for the treatment of waste but forget the current waste problem. None responds to the problem of reintegrating the different outputs into the supply chain, to give it a longer useful life that is the common denominator in the problem of solid waste, only reverse logistics (Table 2).

Transversal pressures were identified to stakeholders, such as: the delay and variability in returns, the unknown quality of the product from the level of consumer information, resistance to change. Also, financial restrictions, and a series of shortcomings (of systems logistics and information, adequate measures of performance, training, education and staff resources, commitment to senior management, rewards, strategic planning and reluctance of merchants, distributors and retailers), among others (Sarache et al. 2017). All these pressures influence in the final structure of reverse logistic network from Tetra Pack packages.

The reverse logistics network stakeholders for the Tetra Pak package are multidimensional and transverse to the stages that make up the supply chain. This result similar to the obtained by Diabat and Govindan (2011), Wu et al. (2012), Sengupta and Mathivathanan (2019). In general, they are the government, suppliers, marketers, distributors, consumers, the market, society, collectors, collectors and sellers. The latter arise in response to institutional, market pressures, shorter product life cycles, greater quality and production process restrictions, among others. Also, due to the growth of a market of the materials that make up the Tetra Pak, the pressure of consumers to take responsibility for the disposal of the waste generated by these containers, the new supplier regulations, the greater competition between industries and the positioning of the Environment as a competitive priority.

According to the results obtained in the characterization survey, educational institutions as schools and universities are a potential source to obtain Tetra Pak wastes. They are generating the Tetra Pak residue, since the consumption of milk, juices is very frequent. In addition, 51% of the students surveyed already know the importance of separation at the source and currently separate this type of waste and 98% are willing to dispose the Tetra Pak package in its special container, located in the educational institution and bring the Tetra packages Pak generated in their dwellings. Finally, the greatest incentive generated to carry out this separation and selective collection is to obtain a bonus in the grades.

References

Battaïa, O., Gupta, S.M.: Reverse supply chains : a source of opportunities and challenges. SME **37**, 587–588 (2015)

Diabat, A., Govindan, K.: An analysis of the drivers affecting the implementation of green supply chain management. Resour. Conserv. Recycl. **55**(6), 659–667 (2011). https://doi.org/10.1016/j.resconrec.2010.12.002

García Olivares, A.: *Recomendaciones táctico-operativas para implementar un programa de logística inversa: Estudio de caso en la industria del reciclaje de plásticos.* (2004) http://www.eumed.net/libros-gratis/2006a/aago/0.htm

Govindan, K., Popiuc, M.N.: Reverse supply chain coordination by revenue sharing contract: a case for the personal computers industry. Eur. J. Oper. Res. **233**(2), 326–336 (2014)

Govindan, K., Soleimani, H., Kannan, D.: Reverse logistics and closed-loop supply chain : a comprehensive review to explore the future. Eur. J. Oper. Res. **240**(3), 603–626 (2015). https://doi.org/10.1016/j.ejor.2014.07.012

Liao, T.Y.: Reverse logistics network design for product recovery and remanufacturing. Appl. Math. Mod. **60**, 145–163 (2018). https://doi.org/10.1016/j.apm.2018.03.003

Mathiyazhagan, K., Haq, A.N.: Analysis of the influential pressures for green supply chain management adoption-an Indian perspective using interpretive structural modeling. Int. J. Adv. Man. Technol. **68**(1–4), 817–833 (2013). https://doi.org/10.1007/s00170-013-4946-5

Mathiyazhagan, K., Sengupta, S., Mathivathanan, D.: Challenges for implementing green concept in sustainable manufacturing: a systematic review. OPSEARCH **56**(1), 32–72 (2019). https://doi.org/10.1007/s12597-019-00359-2

Narayana, S.A., Pati, R.K., Padhi, S.S.: Market dynamics and reverse logistics for sustainability in the Indian Pharmaceuticals industry. J. Clean. Prod. **208**, 968–987 (2019). https://doi.org/10.1016/j.jclepro.2018.10.171

Olugu, E.U., Wong, K.Y., Shaharoun, A.M.: Development of key performance measures for the automobile green supply chain. Resour. Conser. Recycl. **55**(6), 567–579 (2011). https://doi.org/10.1016/j.resconrec.2010.06.003

Panigrahi, S.K., et al.: A strategic initiative for successful reverse logistics management in retail industry. Global Business Rev. **19**(3), S151–S175 (2018). https://doi.org/10.1177/0972150918758096

Pinheiro, E., et al.: How to identify opportunities for improvement in the use of reverse logistics in clothing industries? A case study in a Brazilian cluster. J. Clean. Prod. **210**, 612–619 (2019). https://doi.org/10.1016/j.jclepro.2018.11.024

Qiu, Y., et al.: 'Production routing problems with reverse logistics and remanufacturing. Transport. Res. E-Log. Transport. Rev. **111**, 87–100 (2018). https://doi.org/10.1016/j.tre.2018.01.009

Rachih, H., Mhada, F.Z., Chiheb, R.: Meta-heuristics for reverse logistics: a literature review and perspectives'. Comput. Ind. Eng. **127**, 45–62 (2019). https://doi.org/10.1016/j.cie.2018.11.058

Rubio Lacoba, S., Tomás Bañegil Palacios D Albert Corominas Subias, D. M.: *Empresa: Análisis Y Aplicaciones Universidad De Extremadura Departamento De Economía Aplicada Y Organización De Empresas.* (2003) http://www.pcid.es/public.htm

Sarache, W., et al.: Redesign of a sustainable reverse supply chain under uncertainty : a case study. J. Oper. Res. **151**, 206–217 (2017)

Shaharudin, M.R., Zailani, S., Tan, K.C.: Barriers to product returns and recovery management in a developing country: investigation using multiple methods. J. Clean. Prod. **96**, 220–232 (2015). https://doi.org/10.1016/j.jclepro.2013.12.071

Superintendencia de Servicios Públicos Domiciliarios and Departamento Nacional de Planeación . *Informe Nacional de Disposición Final de Residuos Sólidos.* ISSN: 2422–2941 (2017)

Tetra , P.: *Informe de Sostenibilidad 2017 Claro Colombia.* (2017) https://www.claro.com.co/portal/recursos/co/legal-regulatorio/pdf/Informe_de_sostenibilidad.pdf

Wu, G.C., Ding, J.H., Chen, P.S.: The effects of GSCM drivers and institutional pressures on GSCM practices in Taiwan's textile and apparel industry. Int. J. Prod. Econ. **135**(2), 618–636 (2012). https://doi.org/10.1016/j.ijpe.2011.05.023

Structuring of the Current Status of the Optimization of Production and Logistics Processes in Medium-Sized Companies in the Main Economic Sectors of the Metropolitan Area

Yenny A. Aguirre A.[1](✉), Mauricio Montoya P.[2], Juan E. Pareja P.[3], Edison A. Usma R.[2], Andres Ortega G.[3], and Andres C. Jiménez G.[3]

[1] Coordinación Ingeniería Industrial, Escuela de Ingenierías, Institución Universitaria Salazar y Herrera, Medellín, Colombia
yenny.aguirre@salazaryherrera.edu.co

[2] Ingeniería de Productividad y Calidad, Facultad de Administración, Politécnico Colombiano Jaime Isaza Cadavid, Medellín, Colombia
{mmontoya,edison_usma64141}@elpoli.edu.co

[3] Ingeniería Industrial, Escuela de Ingenierías, Institución Universitaria Salazar y Herrera, Medellín, Colombia
{juan.parejap,andres.ortegag,
andres.jimenezg}@comunidad.iush.edu.co

Abstract. The aim is to know what is the current status of the optimization of production and logistics processes in medium-sized companies in the main economic sectors of the Metropolitan Area. As a result, the potential of the university-company relationship will be enhanced, based on the diagnosis made, to identify which are the main variables to be measured in terms of the optimization of processes by the economic sectors with the greatest potential in the Metropolitan Area, in terms of competitiveness, incursion of professionals in the field, economic contribution to the region, dynamics and tendency. It will be from this that a scenario of what the labor competencies are and, within them, the academic competencies that are indispensable for the careers in the disciplines of productivity and logistics for their curricular content.

Keywords: Prospective · Optimization · University-company · Variables · Competencies

1 Introduction

There have been many different approaches to optimizing business processes in any productive sector; from methodologies of continuous improvement in the early 50s, to Lean Manufacturing tools from the boom in the 80s, to the analysis of information of Big Data in the late 90s. However, a question arises before trying to solve the assorted

M. T. Castañeda Galvis et al. (Eds.): ICIE 2019, LNMUINEN, pp. 11–19, 2021.
https://doi.org/10.1007/978-3-030-49370-7_2

business problems: What is the current state of optimization of production and logistics processes in medium-sized companies in the main economic sectors of the Metropolitan Area? From this same approach, a series of questions arise that, directly or indirectly, link not only to the business sector, but also to academic training, in the attempt to assertively strengthen the skills and profiles of those who will be responsible for that optimization. Some of the other questions that are broken down and will be the subject of the study are:

- What is the scope and impact of processes of optimization in the economic sectors of the Metropolitan Area?
- What are the variables and their respective measurement scales used for process optimization?

2 Methodology

For the methodological development, a series of phases, steps and activities are constructed that allow the achievement of the proposed objectives in a structured way. These phases are:

- In *Phase I (Delimitation)*, a scope of work is defined. That is, what are the criteria for the selection of the economic sectors that will be studied, where it is also possible to argue the interest of medium-sized companies in the Metropolitan Area. Subsequently, a joint methodology of prospective theories is structured that allows the construction of a Scientometric analysis of the more than 50 sources of information, gathering from this a clearer diagnosis on the current state of the optimization of productive and logistic processes, thus delimiting the work system.
- *In Phase II (Definition)*, once the work structure is delimited, the generation of hypotheses of variables and measurement scales begins from a brainstorm for the construction of a list of variables with their respective measurement scales that allows the construction of the measuring instrument in the study. Then, a review of the relationship between variables is made from a list of companies in the sectors and conditions of the selected study universe in order to calculate the sample size, and with them, perform consultancies or field work on the selected companies. Subsequently, a definition of key variables is made based on the recording of results and the use of the MICMAC methodology (input variables, output variables, autonomous variables, objective, key [challenge], determinants, environment, regulators). After that, the construction of a cross-impact matrix is made that allows the definition of final variables and with them the respective report of findings. Finally, the construction of a cross-impact matrix is made that allows the definition of final variables, and with them the respective findings report.
- *In Phase III (Validation)*, the validation with experts begins, for which both the Workshop and the list of experts required for such validation are structured. The SMIC execution then proceeds, which allows the generation of the final report of the meeting, giving way to the structuring of the current state of the optimization of productive and logistic processes objective of this research.

3 Results

3.1 Phase I

For the construction of the state of the art, a search of sources of information was conducted in relation to a series of keywords such as: optimization of processes, economic sectors, productive sectors, cluster, GDP, labor linkage, among others. The objective of this search was to determine which economic sectors have the greatest popularity in terms of process optimization, what are the optimization techniques or models used, what advances have been made in these areas, and what basis will be used to generate a prospective diagnosis. The main results are shown below.

Part of the exercise consists of determining the criteria that will serve as input for the classification of the productive or economic sectors studied, so that the main sectors are selected as sample axes of this research. Below, Table 1 shows the summary of sectors, through which the crossing with the qualification variables can be generated on a basic scale (High, Medium, Low), selecting those with the greatest impact on the High scale.

Table 1. Summary of sectors by study variables

Sector	Subsector	Competitividad	Incursión de profesionales en el medio	Contribución económica a la región	Dinámica y Tendencia	CONTAR
Sector Primario	Agricultura	BAJO	BAJO	ALTO	MEDIO	1
	Insumos agrícolas	BAJO	BAJO	BAJO	BAJO	0
Industria Manufacturera	**Textil y Confección**	**ALTO**	**ALTO**	**ALTO**	**ALTO**	**4**
	Metales, Aceros y derivados	ALTO	ALTO	MEDIO	MEDIO	2
	Caucho y Plástico	**ALTO**	**ALTO**	**MEDIO**	**ALTO**	**3**
	Aseo	ALTO	NA	BAJO	ALTO	2
	Cuero y Marroquinería	ALTO	BAJO	BAJO	ALTO	2
	Alimentos	**ALTO**	**ALTO**	**MEDIO**	**ALTO**	**3**
	Cerámicos	ALTO	BAJO	BAJO	BAJO	1
Servicios	Electricidad, gas, agua	BAJO	MEDIO	ALTO	ALTO	2
	Comercio	**ALTO**	**ALTO**	**ALTO**	**ALTO**	**4**
	Mantenimiento	ALTO	BAJO	ALTO	BAJO	2
	Servicios Sociales	MEDIO	ALTO	ALTO	BAJO	2
	Salud	ALTO	MEDIO	BAJO	ALTO	2

Source: Own Elaboration

In this way, the criteria to be used are defined as follows:

- *Competitiveness*: Represents cooperation scenarios among participants in the sectors. The contribution of productivity, innovation, development, growth, technology and application of strategies are added for increasing the achievement of productive results. The findings of the sources of information were taken as the basis for qualification.
- *Incursion of professionals in the field:* Represents the number of direct or indirect links in the different economic sectors of professionals in the fields of productivity, industrial engineering, logistics and related fields in the metropolitan area. The link to student practices (from industrial engineering, logistics, and productivity and quality programs) of both institutions was taken as a basis for qualification, where 10.1% are linked to companies in the textile and clothing sector, followed by 8% linked to the food sector, 7% linked to the social services sector, 6.6% linked to the commercial sector and 6.4% linked to the rubber and plastic manufacturing sector.
- *Economic contribution to the region:* Represents sales, imports, exports, staff occupation (linking of university professionals) and contributions in general from the sector to the region's economy. The findings from the sources of information were taken as the basis of qualification.
- *Dynamics and trend:* According to the behavior of the markets, aggregate values of the products or services of the sector, adaptability to changes in the environment, compliance with the needs and requirements of the different actors present in the chain. It also refers to the regional and national infrastructure available to a sector to achieve logistics effectiveness and its impact on economic, social and environmental systems as principles of the operation. The participation in Clusters, Prodes and other associative models is taken as the basis of qualification.

In this way, for Phase I, the medium-sized companies of the textile and clothing, rubber and plastic, food and commerce sectors of the Metropolitan Area are defined as the scope for this project, considering the criteria of competitiveness, incursion of professionals in the environment, economic contribution to the region, dynamics and trend.

3.2 Phase II

To define the variables that will be the initial basis of the study, the process is as follows:

- The delimitation was based on the system that was generated based on the Scientometric analysis to create an initial and preliminary list of 20 variables.
- These variables were grouped according to 5 of the 8 existing changes (Defects, Inventories, Reprocesses, Human Talent, Times), as types of deviation in the optimal allocation of resources in the different production and logistics processes.
- Subsequently, this list of variables is sent for validation to a random group of 124 students, entrepreneurs, graduates, teachers and professionals related to the production and logistics processes.
- The respective analysis of results is performed, generating a count of the most representative variables. In addition, extra study variables are obtained in a complementary manner.

Table 2. Results of application of the variables according to consultancies.

Variable	Muda (Tipos de desviación en la asignación óptima de recursos)	Aplica		No Aplica		Total
		N	%	N	%	
Vble 1. Ausentismo en el Trabajo	Talento Humano	25	78%	7	22%	32
Vble 2. Porcentaje de Reprocesos	Reprocesos	23	72%	9	28%	32
Vble 3. Horas Extra	Tiempos	26	81%	6	19%	32
Vble 4. Defectos de Calidad	Defectos	22	69%	10	31%	32
Vble 5. Uso TIC	Talento Humano	26	81%	6	19%	32
Vble 6. Niveles de Inventario	Inventarios	25	78%	7	22%	32
Vble 7. Devoluciones	Defectos	26	81%	6	19%	32
Vble 8. Rotación del Inventario	Inventarios	24	75%	8	25%	32
Vble 9. Trazabilidad de la Operación	Tiempos	26	81%	6	19%	32
Vble 10. Transporte de Mercancía	Tiempos	29	91%	3	9%	32
Vble 11. Rotación del Personal	Talento Humano	27	84%	5	16%	32
Vble 12. I+D+i	Talento Humano	24	75%	8	25%	32
Vble 13. Costos de No Calidad	Reprocesos	23	72%	9	28%	32
Vbles 14. Riesgos e Incertidumbre Operativa	Reprocesos	24	75%	8	25%	32

Source: Own Elaboration

For the calculation of the sample size in relation to how many companies must be diagnosed in order for the study to be representative, there is a universe of 7661 medium-sized companies belonging to the Metropolitan Area according to sources from the Universia Business Guide for 2018. With a probability of occurrence of 0.1 for this finite population, a confidence level of 95% and an error over the estimate of 10%, the sample size, that is, the number of companies in the four sectors of study that should be subject to perform with them the diagnosis of variables, is 34 companies.

The work of consultancies, or field work, was performed in a period of one month. With a total of 70 companies invited to be part of the research, a linkage of 47.14% was achieved, that is, a total of 32 companies participating in the study. 100% was performed through a survey via email. Table 2 shows the result of the application of the study variables by the companies that participated in the consultancy.

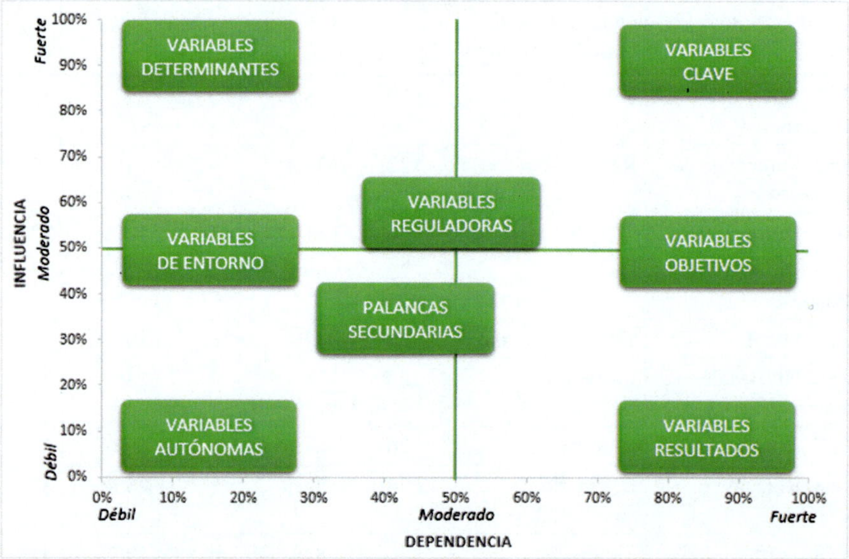

Fig. 1. Summary of sectors by study variables. Source: Own elaboration based on Godet 2007

The variables are projected on a graph of influence and dependence. Its distribution allows identification of seven categories of variables, which have a certain difference in function in the dynamics of the system (Godet 2007). Figure 1 shows these categories. Next, we describe each one of them.

- *Autonomous variables:* They are not a determining part for the future of the system.
- *Objective variables*: These can be influenced so that their evolution is the one desired.
- *Key variables:* They are also called "challenge variables". They disturb the normal functioning of the system.
- *Determining variables:* They become brakes (inhibitors) or motors (propellers) of the system.
- *Environment variables:* The objective is to complement its added value to the system.
- *Regulatory variables:* They become a stopcock to achieve compliance with the key variables.
- *Variable results:* These are variables that cannot be approached head on, but rather through the dependent variables in the system.
- *Secondary Levers:* Complementary to the Regulatory Variables and allow their evolution towards the key variables.

According to the consultancies performed, the 14 study variables were categorized by the companies into 4 groups: regulatory, objective, environment and secondary lever as shown in Table 3.

Table 3. Results of application of the variables according to consultancies

Variable	Dependencia	Influencia	Categoría
Vble 1. Ausentismo en el Trabajo	64%	64%	Reguladora
Vble 2. Porcentaje de Reprocesos	61%	52%	Objetivo
Vble 3. Horas Extra	46%	50%	Entorno
Vble 4. Defectos de Calidad	55%	59%	Reguladora
Vble 5. Uso TIC	58%	58%	Reguladora
Vble 6. Niveles de Inventario	56%	48%	Palanca Secundaria
Vble 7. Devoluciones	46%	46%	Entorno
Vble 8. Rotación del Inventario	42%	54%	Entorno
Vble 9. Trazabilidad de la Operación	50%	65%	Reguladora
Vble 10. Transporte de Mercancía	52%	52%	Palanca Secundaria
Vble 11. Rotación del Personal	52%	52%	Palanca Secundaria
Vble 12. I+D+i	58%	50%	Objetivo
Vble 13. Costos de No Calidad	65%	65%	Reguladora
Vble 14. Riesgos e Incertidumbre Operativa	54%	50%	Palanca Secundaria

Source: Own elaboration

In this way, with Phase II, the resulting variables in the study with the highest relevance for the measurement of the optimization of the productive and logistics processes are: absenteeism at work, percentage of reprocessing, overtime, quality defects, use of tic, inventory levels, returns, inventory turnover, non-quality costs, transportation of materials, staff turnover, traceability of the operation, Research and Development, risks and operational uncertainty; categorized according to the 32 companies of the study in variables of environment, regulators, secondary lever and objective.

3.3 Phase III

For the validation phase, 19 experts participated in the Workshop, who followed the following work method: There was a Research Leader per variable table (four tables, in total). The experts rotated randomly through all four tables (with the aim of creating a diversity of concepts as the exercise progressed, thanks to the positions and arguments focused on in relation to the variables addressed). The rotation was done every 7 min, reading the guide of each Variable table. It was essential to highlight the ideas, extra variables, relocation of variables (if necessary), measurement scales and other opinions using the material available at each table. The Investigative Leader of each table at the end of the four rotations gave the final conclusions of the exercise in a unified way in the Matrix of Cross Impacts.

The Workshop exercise lasted two hours, with positive acceptance and of great interest to the attendees. Below in Fig. 2, the results are shown in a comparative manner. In the graph on the left, the location of the 14 variables in the four categories is shown according to the consultancies previously performed. The graph on the right shows the final location of the 14 variables (plus two extra resulting from the exercise) after the validation by the experts.

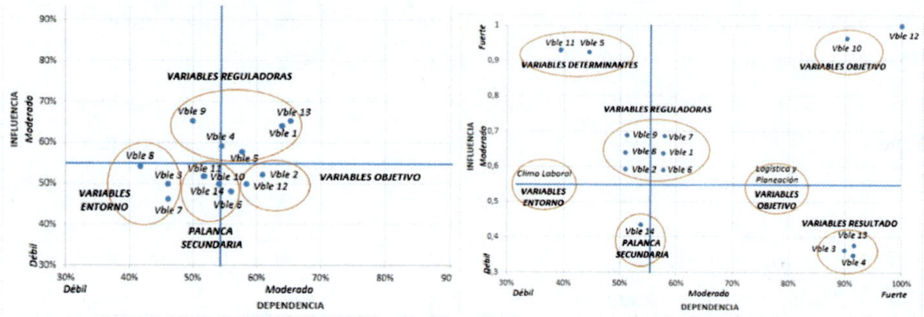

Fig. 2. Summary of sectors by study variables Source: Own elaboration

Table 4. Results of application of the variables according to consultancies

Variable	Dependencia	Influencia	Categoría	Workshop
Vble 1. Ausentismo en el Trabajo	64%	64%	Reguladora	Reguladora
Vble 2. Porcentaje de Reprocesos	61%	52%	Objetivo	Reguladora
Vble 3. Horas Extra	46%	50%	Entorno	Resultado
Vble 4. Defectos de Calidad	55%	59%	Reguladora	Resultado
Vble 5. Uso TIC	58%	58%	Reguladora	Determinantes
Vble 6. Niveles de Inventario	56%	48%	Palanca Secundaria	Reguladora
Vble 7. Devoluciones	46%	46%	Entorno	Reguladora
Vble 8. Rotación del Inventario	42%	54%	Entorno	Reguladora
Vble 9. Trazabilidad de la Operación	50%	65%	Reguladora	Reguladora
Vble 10. Transporte de Mercancía	52%	52%	Palanca Secundaria	Clave
Vble 11. Rotación del Personal	52%	52%	Palanca Secundaria	Determinantes
Vble 12. I+D+i	58%	50%	Objetivo	–
Vble 13. Costos de No Calidad	65%	65%	Reguladora	Resultado
Vble 14. Riesgos e Incertidumbre Operativa	54%	50%	Palanca Secundaria	Palanca Secundaria

Source: Own elaboration

In this way, with Phase III, in the Workshop of experts, there is success in the categorization of variables of 21%. 71% changed categories, and 7% required a revision (as shown in Table 4). In addition, two more study variables were found: work environment, and logistics and planning.

4 Discussion

It is evident after the research conducted that the business sector, in matters of optimization of production and logistics processes, lacks a clear definition of variables with their respective measurement scales, and therefore the impact they can generate on their competitive performance.

The structuring of business consultations, based on previously defined measurement and study variables, will serve as a model of articulation between the university and the company. On the other hand, these types of exercises will also become diagnostic tools for the companies participating in the research study in their decision-making process of productivity, logistics and competitiveness, thus successively generating prosperity on the measurement variables selected for the optimization of production and logistics processes.

The requirement of a characterization of the business environment and the relevance of the academic competencies recommended with respect to those required by the business sector in matters of optimization are generated as the final scenario.

References

Godet, M., Durance, P: La caja de herramientas de la prospectiva estratégica. Problemas y métodos. Cuadernos de LIPSOR, Donostia-San Sebastián, España (2007)

Supply Chain

Effects of Knowledge Management and Sustainable Supply Chain Management on the Sustainable Performance of the Industrial Manufacturing Sector

Bertha Lucía Santos Hernández(✉), Sandra Yesenia Pinzón Castro, and Elena Patricia Mojica Carrillo

Centro de Ciencias Económicas y Administrativas, Universidad Autónoma de Aguascalientes, Aguascalientes, Mexico
Lucy_santoshdz@yahoo.com.mx, {sandra.pinzon, emojica}@edu.uaa.mx

Abstract. The quantity and constant flow of material, financial, technological, information resources and diverse interests of stakeholders make the management of supply chains more complex, so that the inclusion of the principles of Sustainable Development together with the management of the volume of knowledge that flows along the supply chain represent a real challenge; even more so, when it must be demonstrated that the effort employed in the practices of knowledge management and sustainability are reflected in economic, environmental and social performance. Therefore, the following questions arise: What is the outcome of sustainability and knowledge management practices in economic, environmental and social performance? How are knowledge management and sustainability management constructs related in the industrial manufacturing context? What aspects should organizations strengthen to achieve sustainability in all its dimensions? The research made it possible to identify some aspects of knowledge management that need to be strengthened in order to achieve the objectives of sustainable development, the exposition of current practices in matters of economic, environmental and social sustainability, and the dimensions of sustainable performance that are most impacted were identified.

Keywords: Sustainability · Supply chain · Knowledge management · Sustainable performance

1 Introduction

Economic development is a priority for many countries; however, it cannot be possible without addressing the problems faced by our environment, particularly developing countries where poverty, inequality and damage to the environment seem to characterize them; in order to eradicate them, the Sustainable Development Objectives were proposed in Agenda 2030, the achievement of which will be possible with cooperation and coordination of regional policies from the context of each country and with the

M. T. Castañeda Galvis et al. (Eds.): ICIE 2019, LNMUINEN, pp. 23–30, 2021.
https://doi.org/10.1007/978-3-030-49370-7_3

participation of all sectors: governments, non-governmental organizations, civil society and the business community (Álvarez 2016). However, the results at the global level reveal that there is still a need to intensify efforts so that the principles of sustainability are really adopted, become the central axis of the strategies and change the focus in the performance of the industry (Turker and Altuntas 2014).

In this sense, the industrial manufacturing sector represents an area of opportunity to have a positive impact on the environment through the analysis of the life cycle of the product, making it possible to design products that respect the environment (Cappa et al. 2016). As a result, sustainability strategies must be integrated throughout the organization and deployed to all areas and companies within their chains nationally and even beyond their borders (Turker and Altuntas 2014).

In order to achieve this, it is indispensable to have mechanisms to improve their knowledge in its internal and external dimensions, that is, through formal and informal training programs, incentives and performance evaluations for both their own employees and their suppliers (Andersen and Skjoett-Larsen 2009). It is believed that interest in knowledge management is currently increasing, and Latin American organizations are adopting it, which will increase the intellectual capital of each country (Liberona and Ruiz 2013).

Thus, knowledge management becomes an influential factor in managing the supply chain (Jamalian et al. 2018) and as a background to sustainable supply chain practices (Abd Rahman et al. 2016); however, although knowledge management and its link to sustainable supply chain management has generated interest in the academic community its relationship has been poorly addressed and most of the research found tends to focus on the environmental aspect; consequently, holistic research integrating economic, environmental and social dimensions is needed (Dao et al. 2011); more research is also needed that considers the characteristics of emerging markets (Zhou et al. 2016); therefore, this study aims to fill these gaps by determining the relationship between knowledge management and sustainable supply chain practices and their influence on sustainable performance in its economic, environmental and social dimension in an entity belonging to an emerging economy country.

2 Methodology

Knowledge management has been defined as the strategic and systematic management of knowledge immersed in an organization through its storage, distribution and application in order to create value (Lundvall and Johnson 1994); it has the potential to transform knowledge into a competitive advantage of the business (Arvin et al. 2014); it involves in its management processes such as the acquisition, conversion, application and protection of knowledge (Abd Rahman et al. 2016) and within its management the human factor stands out, which has become the main productive resource since it is the holder of the knowledge that must be managed together with the identification and evaluation of the knowledge and good practices that the organization has to become itself into better products and services (Ángel Oteo 2003).

The review of the literature made it possible to identify some problems related to knowledge management; for example, to have all the pertinent characteristics of documentation; the use of advanced support tools that speed up decision-making; the lack of knowledge of the dimensions of the intellectual capital of organizations; procedure manuals are not consulted or are not updated; the lack of knowledge management methods, which makes it necessary to have the advice of consultants or universities that can promote and teach the best practices of knowledge management; but, above all, it is necessary to have a knowledge management system in place requires the support of management whose direction and leadership are important in implementation. (Liberona and Ruiz 2013).

On the other hand, sustainable supply chain management is interpreted as the design, organization, coordination and control of supply chains aimed at achieving sustainability (Pagell and Shevchenko 2014), its management implies maintaining economic benefits without damaging the environment and society (Abd Rahman et al. 2016). Sustainable practices allow companies to perform better in the market and therefore perform better financially; in turn, having resources allows for environmental marketing strategies where tangible and intangible assets make the difference with respect to competition; they also reduce some costs such as those related to energy consumption, the responsible use of materials and inputs, processes are more efficient and an improved image or reputation is acquired; in short, there is a competitive advantage that has an impact on the market and on clients who are increasingly aware of caring for the environment (Leonidou et al. 2013).

Figure 1 shows the results of the search for publications on Knowledge Management and Sustainable Supply Chain Management in the area of Negotiation, Administration, Accounting and Marketing; however, there was not enough evidence in the Elsevier database to relate them holistically, i.e. to integrate the environmental, social and economic dimensions (Das 2017); however, there are some indications of this relationship in research on knowledge management and social responsibility, these are widely used by organizations within the processes of administration; however, on many occasions they are not linked within the strategic management, but reflects an increased interest in finding options that allow them to be more competitive through strategies that have been used in the international context (Saldarriaga Ríos 2013).

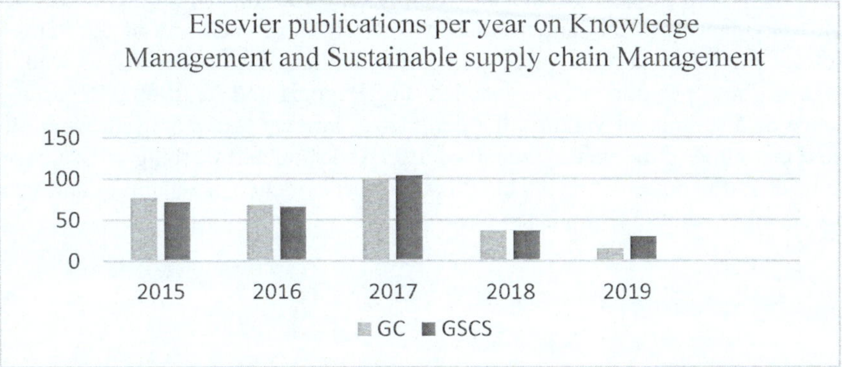

Fig. 1. Publications on knowledge management and supply chain sustainability management located in Elsevier in the last five years.

In this way, it is possible to point out that integrating knowledge and sustainable practices at the supply chain can increase the company's profits and constitute its competitive advantage by emphasizing the importance of profits, people and the planet, in accordance with the vision of the Triple Bottom Line TBL which has a positive impact on the sustainability performance of companies (Dao et al. 2011). Given the above, the research hypotheses are presented in the following manner:

H1: Knowledge management has positive, direct and significant effects on sustainable supply chain practices.

H2: Sustainable supply chain practices have a direct, positive and significant effect on the sustainable performance of the supply chain management.

For verification, empirical research was designed with a quantitative approach, non-experimental research design, transversal, causal with explanatory scope; the sample size was obtained by simple random sampling, the sample is made up of 224 small, medium and large industrial manufacturing companies in a city of Mexico; information was collected through the perception of managers, quality managers, production or logistics of companies with more than ten employees; since their position influences the implementation of this type of strategies in organizations and, therefore, in their performance. The information was collected through a questionnaire integrated by standardized instruments; to measure knowledge management, the instrument developed by Bozbura (2007) integrates the following dimensions: training and mentoring of employees, knowledge management policies and strategies, acquisition of external knowledge, and influence of the organizational culture; to measure sustainable management (practices and performance) of the supply chain, an adaptation of Das (2017) was used.

The factorial analysis was used to evaluate the reliability and validity of the measurement scales; reliability was measured using Cronbach's Alfa, the Composite Reliability Index (IFC) and the average of the variance extracted (AVE); Cronbach's Alfa values exceeded 0.7; factorial loads exceeded 0.6 (Hair et al. 1995); the value obtained from IFC and IVE respectively complies with the parameters indicated by Fornell and Larcker (1981) as can be observed in Table 1. The indices NFI = 0.750, NNFI = 0.859, CFI = 0.875, RMSEA = 0.057 show good fit in the model once the items with factorial loads that did not comply with the indicated criteria were eliminated; to evidence the convergent validity, the values of each one of the items were significant $p < 0.001$ and the values of the standardized factorial loads, as well as the average of these per dimension exceeded 0.6 (Bagozzi and Yi 1988). Regarding the evidence of discriminant validity, the confidence interval test was used, where it was verified that none of the values contained 1.0 (Anderson and Gerbing 1988); based on these results, it is possible to conclude that there is evidence of reliability and validity, both convergent and discriminant.

Table 1. Internal consistency reliability and convergent validity.

Variable	Indicators	Standarized coefficient	t value	Cronbach´s Alpha	CR	AVE
Training and mentoring employees	TME1	0.722	1	0.836	0.844	0.576
	TME2	0.822	11.749			
	TME3	0.785	11.147			
	TME4	0.701	10.019			
	Mean	0.7575				
Policies and strategies of knowledge management	PS1	0.704	1	0.89	0.894	0.678
	PS2	0.776	10.811			
	PS3	0.797	11.1			
	PS4	0.834	11.578			
	PS5	0.846	11.738			
	Mean	0.7914				
Knowledge capturing and acqisition from outside	KCA1	0.672	1	0.835	0.838	0.567
	KCA3	0.791	10.789			
	KCA4	0.656	8.698			
	KCA5	0.872	11.287			
	Mean	0.74775				
Effect of organizational culture	EOC2	0.84	1	0.67	0.823	0.612
	EOC3	0.858	7.712			
	EOC4	0.629	7.641			
	Mean	0.7756667	6.635			
Environmental management practices (EMP)	EMP1	0.89	1	0.925	0.927	0.681
	EMP2	0.911	20.934			
	EMP3	0.892	20.024			
	EMP4	0.819	16.674			
	EMP5	0.767	14.777			
	EMP6	0.641	11.185			
	Mean	0.82				
Socially inclusive practices for employees (SPE)	SPE1	0.682	1	0.817	0.831	0.623
	SPE2	0.856	10.618			
	SPE3	0.82	10.387			
	Mean	0.786				
Socially inclusive practices for community (SPC)	SPC2	0.803	1	0.86	0.867	0.766
	SPC2	0.942	2.878			
	Mean	0.8725				
Operations practices (OP)	OP1	0.831	1	0.873	0.874	0.585
	OP2	0.732	12.453			
	OP3	0.628	10.203			
	OP4	0.843	15.464			
	OP5	0.769	13.371			
	Mean	0.7606				
Environmental performance (EPR)	EPR1	0.901	1	0.882	0.886	0.664
	EPR2	0.918	19.428			
	EPR3	0.746	14.276			
	EPR5	0.668	11.973			
	Mean	0.80825				
Employee-centred social performance (ESP)	ESP1	0.747	1	0.784	0.786	0.551
	ESP2	0.738	9.695			
	ESP3	0.741	9.756			
	Mean	0.742				
Community-centred social performance (CSP)	CSP1	0.629	1	0.843	0.855	0.669
	CSP2	0.91	10.497			
	CSP3	0.885	10.395			
	Mean	0.808				
Financial performance	FP1	0.76	1	0.92	0.921	0.625
	FP2	0.757	11.698			
	FP3	0.797	12.422			
	FP4	0.777	12.054			
	FP5	0.762	11.793			
	FP6	0.847	13.334			
	FP7	0.827	12.963			
	Mean	0.7895714				

Source: Own contribution from results obtained.

3 Results

A confirmatory factorial analysis was made to the theoretical model through the Structural Equations Model (SEM) using the EQS 6.1 software (Bentler 2006) with which it was possible to check the structure of the model and contrast the hypotheses; the nomological validity was analyzed through the performance of the Chi-square where the measurement obtained from the theoretical model and the adjusted model was compared (Fornell et al. 1988). Table 2 shows the values obtained for hypothesis 1 where it is indicated that knowledge management directly and positively influences sustainable supply chain management practices with moderate significance; as for Hypothesis 2, it was found that direct and positive influence was significant with sustainable supply chain performance. It is therefore concluded that knowledge management positively influences sustainable supply chain practices and these, in turn, positively influences sustainable supply chain performance.

Table 2. SEM results of the structural model

Hypotheses	Path	Standarized coefficient	t value	p value
H1: Knowledge management has positive, direct and significant effects on sustainable supply chain practices	KM → SSC practices	0.431	29.469	0.000
H2: Sustainable supply chain practices have a direct, positive and significant effect on the sustainable performance of the supply chain management	SSCP → SSC performance	0.651	35.003	0.000

Source: Own contribution from results obtained.

4 Discussion

Companies consider knowledge as a valuable strategic element that influences competitiveness; however, the descriptive analysis shows little interest in acquiring knowledge from outside the company, of its environment, and in working together with its partners in the supply chain, which could speed up its response rate to the challenges that the environment poses every day; therefore, knowledge management should become a priority which apparently has not yet been achieved; its benefits would be reflected in the improvement of their organizational communication, problem solving processes, decision making, innovation of their products or processes, efficiency, customer service and participation in the market; this would allow them to be more competitive and consequently contribute to economic development (Torres et al. 2014).

The theory indicates that small and medium enterprises tend to be more responsible with their environment in comparison with microenterprises, the separation of ownership and corporate control favor the development of socially responsible practices in small and medium enterprises. As for the most important aspects of the environmental dimension, waste management is the factor that has the greatest impact on financial performance; however, no evidence has been identified of detailed analysis of the inclusion of the principles of sustainable development in supply chain management in small and medium enterprises, what has been found is that they are more involved with the environmental and social practices of the supply chain (Kot 2018). However, the values obtained contradict this assertion since small companies reflected a lower level of implementation of sustainable practices compared to medium and large companies, especially in the social dimension. Another limitation corresponds to the analysis of financial performance, which was obtained through the survey, but could be complemented with the analysis of financial statements; however, given that these are confidential data, it is unlikely to obtain them. Regarding future lines of research, it is advisable to expand the sample size, include other sectors of the industry and make comparative analyses with other geographical contexts.

Compliance with Ethical Standards. Ethical Approval: This research was approved by the Campus committee. All procedures performed involving human participants were in accordance with the ethical standards of the institutional research committee and its later amendments or comparable ethical standards.

Informed Consent: it was obtained from all individual participants included in the study, all participants were informed of the objective of the study and agreed to participate in this research by answering the questionnaire voluntarily, All information taken from the study was coded to protect each subject's name. No names or other identifying information was used when discussing or reporting data and data collected is in a secured place.

Conflicts of Interest: The authors report no conflicts of interest.

References

Abd Rahman, A.A., Tay, M.Y., Aziz, Y.A.: Potential of knowledge management as antecedence of sustainable supply chain management practices. Int. J. Supply Chain Manag. **5**(2), 43–50 (2016)

Álvarez, A.M.: Retos de América Latina: Agenda para el Desarrollo Sostenible y Negociaciones del siglo xxi. Probl. desarrollo **47**(186), 9–30 (2016). https://www.sciencedirect.com/science/article/pii/S0301703616300219

Andersen, M., Skjoett-Larsen, T.: Corporate social responsibility in global supply chains. Supply Chain Manag.: Int. J. **14**(2), 75–86 (2009). https://doi.org/10.1108/13598540910941948

Anderson, J.C., Gerbing, D.W.: Structural equation modeling in practice: a review and recommended two-step approach. Psychol. Bull. **103**(3), 411 (1988)

Ángel Oteo, L.: La gestión del conocimiento: un paradigma sostenible. Rev. Calidad Asistencial **18**(3), 139–140 (2003). https://doi.org/10.1016/s1134-282x(03)77590-3

Arvin, M., Akbari, M., Moghimnejad, M.: The study of various models of knowledge management. Kuwait Chap. Arab. J. Bus. Manag. Rev. **33**(2538), 1–12 (2014)

Bagozzi, R.P., Yi, Y.: On the evaluation of structural equation models. J. Acad. Mark. Sci. **16**(1), 74–94 (1988)

Bentler, P.M.: EQS 6 structural equations program manual, Los Angeles: BMDP Statistic Software (2006). http://www.econ.upf.edu/~satorra/CourseSEMVienna2010/EQSManual.pdf

Bozbura, F.: Knowledge management practices in Turkish SMEs. J. Enterp. Inf. Manag. **20**(2), 209–221 (2007)

Cappa, F., et al.: How to deliver open sustainable innovation: an integrated approach for a sustainable marketable product. Sustainability (Switzerland) **8**(12), 1341 (2016). https://doi.org/10.3390/su8121341

Dao, V., Langella, I., Carbo, J.: From green to sustainability: information technology and an integrated sustainability framework. J. Strategic Inf. Syst. **20**, 63–79 (2011). https://doi.org/10.1016/j.jsis.2011.01.002

Das, D.: Development and validation of a scale for measuring sustainable supply chain management practices and performance. J. Clean. Prod. **164**, 1344–1362 (2017). https://doi.org/10.1016/j.jclepro.2017.07.006

Fornell, C., Larcker, D.F.: Evaluating structural equation models with unobservable variables and measurement error. J. Mark. Res. **18**(1), 39–50 (1981)

Fornell, C., et al.: Anderson and Gerbing 1988.pdf. **103**(3), 411–423 (1988)

Hair, J.F., Anderson, R.E., Tatham, R.L., Black, W.C.: Multivariate Date Analysis with Readings. Prentce, Englewood Cliff (1995)

Jamalian, A., Ghadikolaei, A.S., Zarei, M., Ghasemi, R.: Sustainable supplier selection by way of managing knowledge: a case of the automotive industry. Int. J. Intell. Enterp. **5**(1–2), 125–140 (2018)

Kot, S.: Sustainable supply chain management in small and medium enterprises. Sustainability (Switzerland) **10**(4), 1–19 (2018). https://doi.org/10.3390/su10041143

Leonidou, L.C., Leonidou, C.N., Fotiadis, T., Zeriti, A.: Resources and capabilities as drivers of hotel environmental marketing strategy: implications for competitive advantage and performance. Tour. Manag. **35**, 94–110 (2013). https://doi.org/10.1016/j.tourman.2012.06.003

Liberona, D., Ruiz, M.: Análisis de la implementación de programas de gestión del conocimiento en las empresas chilenas. Estudios Gerenciales. Univ. ICESI **29**(127), 151–160 (2013). https://doi.org/10.1016/j.estger.2013.05.003

Lundvall, B.Ä., Johnson, B.: The learning economy. J. Ind. Stud. **1**(2), 23–42 (1994)

Pagell, M., Shevchenko, A.: Why research in sustainable supply chain management should have no future. J. Supply Chain Manag. **50**(1), 44–55 (2014)

Saldarriaga Ríos, J.G.: Responsabilidad social y gestión del conocimiento como estrategias de gestión humana1. Estudios Gerenciales. Elsevier **29**(126), 110–117 (2013). https://doi.org/10.1016/s0123-5923(13)70026-3

Torres, C.A.B., González, J.S.F., Arango, H.D.S.: Importancia de la gerencia del conocimiento: contrastes entre la teoría y la evidencia empírica. Estudios Gerenciales **30**(130), 65–72 (2014). http://www.icesi.edu.co/revistas/index.php/estudios_gerenciales/article/view/1766/ html

Turker, D., Altuntas, C.: Sustainable supply chain management in the fast fashion industry: an analysis of corporate reports. Eur. Manag. J. **32**(5), 837–849 (2014). https://doi.org/10.1016/j.emj.2014.02.001

Zhou, K.Z., et al.: Supply chain management in emerging markets. J. Oper. Manag. **46**, 1–4 (2016). https://doi.org/10.1016/j.jom.2016.07.007

Enterprise Architecture for Analysis of Supply Chain in the Hotel and Tourism Sector

Alexander Aragón Ch.[1(✉)], Mónica Patricia Sarria Y.[2(✉)], and Maritza Correa-Valencia[1(✉)]

[1] Faculty of Engineering, Department of Operations and Systems, Universidad Autónoma de Occidente (UAO), Cali, Colombia
{aaragon,mcorrea}@uao.edu.co

[2] Faculty of Engineering, Industrial Engineering Program, Fundación Universitaria Católica Lumen Gentium (Unicatólica), Cali, Colombia
msarria@unicatolica.edu.co

Abstract. The Lean Six Sigma process improvement model developed for the hotel and tourism sector involves, in addition to new industrial engineering tools applied in a timely manner to a key internal process, provides the entrepreneur with a holistic view of the supply chain in which their business is framed. Based on the review of Lean Six Sigma literature, as well as the concepts of Integration and Enterprise Architecture, a new Six Sigma model is proposed; it incorporates a Canvas Business Model and Process Modeling concepts. These provided a broader picture regarding Intra and Inter-company operations and provided the respective opportunities for improvement that will be considered later in the project's progress.

Keywords: Supply chains · Lean Six Sigma · Business integration · Enterprise Architecture · Process Modeling

1 Introduction

Enterprise architecture identifies the main components of the organization and their relation to achieve business objectives. It acts as an integrating force between aspects of business planning, operation, technology, etc. (Urbaczewski and Mrdalj 2006). It has been identified that apart from shortcomings in its internal processes, there are other factors related to suppliers and the distribution of the finished product, which impact operations and, consequently, the competitiveness of SMEs (Fonseca et al. 2018).

Incorporating the DMAIC methodology - Define, Measure, Analyze, Improve, Control (Pyzdek and Keller 2010), provides an orderly and well-structured way to address common problems in small and medium enterprises. However, it focuses mainly on internal (intra-company) processes, leaving aside inter-company relationships along the supply chain. This is why including other industrial engineering tools is key to extending the scope of the methodology.

The experience acquired previously in the research group was considered for the configuration of a model that incorporates Lean Six Sigma (LSS) tools for application

M. T. Castañeda Galvis et al. (Eds.): ICIE 2019, LNMUINEN, pp. 31–38, 2021.
https://doi.org/10.1007/978-3-030-49370-7_4

in the service sector, specifically in hotel and tourism, considering the fact that Cali has been categorized since 2018 as a Special, Sports, Cultural, Tourist, Business and Services District (Colombia 2018).

Through the incorporation of Supply Chain concepts, the *MejorTour* Project[1] was created to allow businessmen to extend the scope of improvement strategies. This is because it has been found that SMEs in the hotel and tourism sector in Cali do not have the production conditions and linkages needed to strengthen their participation in the local market and compete in national and international markets, especially now that Santiago de Cali is established as a Special, Sports, Cultural, Tourism, Business and Services District. According to the Tourist Information Center of Colombia CITUR, Valle del Cauca presents a very slight growth from 2010 to 2017 in comparison with other regions of the country.

To address this issue, LSS, defined as a synergistic process that creates a value flow map of the process which identifies the costs of adding or not adding value (Voehl 2013), and captures the voice of the customer to define the critical quality problems that affect it, will be used. The projects within the improvement process are prioritized according to the time that the results take. This prioritization process inevitably identifies activities with high defect rates - Six Sigma (SS) tools or downtime - Lean Manufacturing (LM) tools, by adopting the DMAIC methodology.

Something LM and SS have in common is that they arise as philosophies after the Second World War as an effort to improve the production processes of companies. Toyota Company and Motorola become managers of these philosophies (Ohno 1988; Sarria et al. 2017) in an attempt to increase the quality and profitability of their companies. Over the years these philosophies evolved, merged and gave way to what is now known as LSS. In it, according to (Pérez-López and García-Cerdas 2014) the focus on the waste and restructuring of processes, characteristic of LM, is used to make them more efficient, fast and agile when responding to customer needs as well as the precise focus of SS where data is used to understand the behavior of processes and identify improvements. Each of these approaches makes a great contribution to continuous improvement from different fronts; the reduction of defects in a structured way in the case of SS and the improvement of process and information flow through Lean Manufacturing or LM (Jiju Antony 2016).

LSS makes use of the different visual tools of LM and statistical tools of SS and their implementation is done through the DMAIC methodology. Through five stages, it achieves the objectives of process improvement. (Pramod Kumar 2017), mentions that in the stage *To Define*, each process involved in the project work area is known. In the stage *To Measure*, one or more critical characteristics are selected, necessary measures are taken, the results are recorded, the measurement systems are evaluated, and the short-term capacity of the process is estimated. In the stage *To Analyze*, the collected information is analyzed and the factors that allow for a substantial improvement and an increase in process performance are identified. In the stage *To Improve*, improvement strategies that seek the optimal performance of the process are implemented. Finally, in

[1] Improvement of processes with Lean Six Sigma, to increase the competitiveness of companies in the hotel and tourism sector linked to the Tourist Board of the Multisectorial Group in Cali. Universidad Autónoma de Occidente and Fundación Universitaria Católica Lumen Gentium.

the stage *To Control*, once the achievement of improvement objectives has been validated, tools are designed to monitor the process and control that the improvements are sustained over time.

Within the objectives of the research project, one of the most important and essential is to diagnose the current status of 20 companies linked to the Tourist Board of Cali, taken as a pilot sample, through the characterization and modeling of their mission processes with a focus on supply chains within the framework of university-industry collaboration, as centers of thought and regional development, contributing to the sustainable development of businesses in the region.

2 Methodology

The methodological structure of the MejorTour project is proposed in 5 phases:

(i) Literature review, focused on the application of LM, SS or LSS in services; mainly on issues such as value creation in tourism services in the subsectors of accommodation, restaurants, transport and tourism services (parks, recreation, tourist destinations and operators).
(ii) Diagnosis, which coincides with the Definition stage of DMAIC, in which the pilot companies are selected and the missionary processes of each one are modeled.
(iii) Model LSS. This involves the Measuring and Analyzing stages of DMAIC. According to the improvement opportunities identified, the LSS tools to be applied are selected, and the necessary interventions and training are planned for all parties involved in the project.
(iv) The development phase of the university-business collaborative model is carried out in parallel throughout the advance of the project. This proposal accomplishes the articulation of those responsible for providing relevant information in companies, as well as students and researchers in universities, all mediated by the use of ICT.
(v) Validation of the model, corresponding to the Improving and Controlling stage of DMAIC. Here, companies are intervened, performance indicators are measured again to verify improvements, and training is used to ensure that changes last over time. Figure 1 presents a summary table with the methodological design of the project.

3 Results

3.1 Literature Review

For the development of phase 1, 89 documents were analyzed according to Table 1, 51 of which were selected. In 55% of the cases, the authors use SS tools, 20% LM and 14% LSS, which ratifies the importance of using the DMAIC methodology proper to these philosophies. Likewise, it was found that, exclusively related to tourism services, 10% of the authors applied Value Creation tools (CV), 8% LM tools, 33% SS tools, 6% LSS tools and 2% applied the three methodologies (LM, SS and LSS).

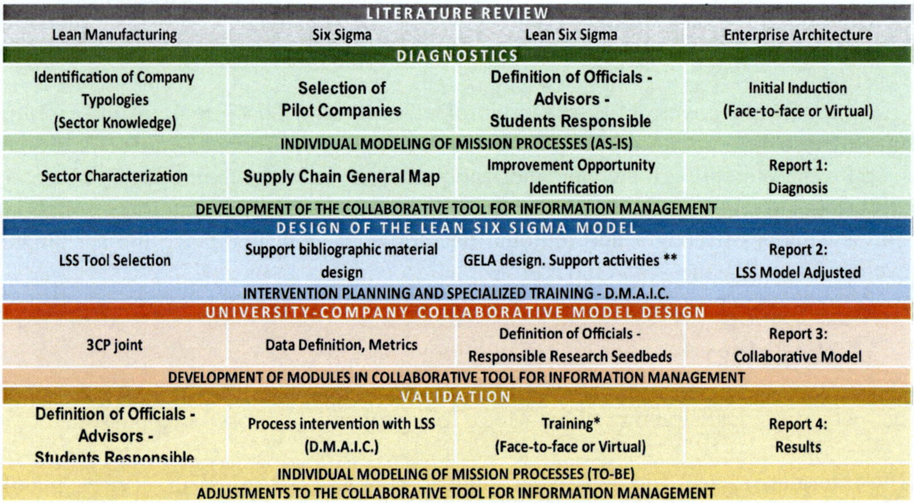

Fig. 1. MejorTour project methodological design. (* The interactive classrooms of the UAO will be used as much as possible for virtual or face-to-face training on Campus. Support from the Study Group on Applied Ludic - GELA. ** GELA participation, for the development of practical training activities.)

Table 1. State of the art composition

Search window		Analyzed documents			Selected documents	
2001–2019		89			51	
>5 years 37%	<5 years 63%	5% theses	11% papers	84% scientific articles	General services 41%	Service in tourism 59%

According to the findings, the tools used in tourism services in the Definition stage are surveys and interviews, with 43% of use by the authors. 20% preferred the project scope statement of improvement and measurement of service quality. Additionally, in 17% of the cases the SIPOC diagram is used (Suppliers, Inputs, Process, Outputs, and Customers) and process maps with 13%.

The results of the bibliographic research regarding the tourism services sector showed that the Canvas business model is used, supported by a more structured process model, through BPMN notation (Business Process Model and Notation). These are tools considered for the development of the MejorTour project.

3.2 Diagnosis

Regarding the supply chain management, the "Definition" stage is adjusted to use the Canvas Business Model (Robinson and Lock 2014; Osterwalder et al. 2005), as a key information source, related to the 9 building blocks of the model: Key Partners (Suppliers), Key Activities (Processes), Key Resources, Value Proposition (Products/Services), Customer Relationships, Channels, Customer Segments, Cost Structure and Revenue Sources. This type of representation also allows building a general company SIPOC diagram, which is then detailed with the selection of a particular process.

As for the integration model and enterprise architecture, the Zachman Architecture reference framework will be followed; it has a two-dimensional structure, containing six rows (Transformation) and six columns (with communication questions), to classify and organize the elements of a company (Goethals 2003).

The MejorTour project is currently under development, the Definition stage has already been completed in 16 of the 20 companies. This process is accompanied by the development of a collaborative information management tool in which all the data supplied has been incorporated (https://sites.google.com/uao.edu.co/mejortour/dmaic).

As relevant findings from the construction of the Canvas Business Model incorporated in it, represented as a supply chain in Fig. 2, where the information provided by the 16 companies was gathered, it was possible to verify that they depend to a large extent on third parties for the provision of their services, such as transport, telecommunications, specialized personnel, and food. However, other possible sources of income such as the sale of advertising, inputs, food or financial investments are under exploited.

With regard to customers, it should be noted that communication with them is principally through social networks, which are their main channels of contact and means of relationship, although they do not have adequate resources and processes to guarantee their loyalty.

With reference to the internal supply chain, the absence of function manuals and process documentation, as well as training programs for human talent, is noteworthy.

As additional information, Fig. 3 shows the grade given by entrepreneurs to their suppliers or partners of key processes for the company. There are 4 key processes identified as security, telecommunications, maintenance, and legal and financial advice, where the service provided has been rated as very good or acceptable.

As for the architecture of the University-Company collaboration model, Fig. 4 shows the summary of the proposal where the interested parties are detailed with their roles as views, models and stages, in an adaptation of the Zachman Architecture Reference Framework (Zachman 2008). It specifies the motivation that corresponds to the Why; the data from where it arises related to the What; the function of the actors that is associated to the How; the means of permanent contact or Where; the people that will be involved or Who; and finally, the time of execution that belongs to When.

It is worth noting that the main focus of architecture centers on research projects, as the interested Designer of the requirements necessary for collaboration, because, from their Logical view of its operation, the Systems Model that will represent what is intended to achieve in the University-Company collaboration is designed and will be developed by the Builders and Integrators in the architecture.

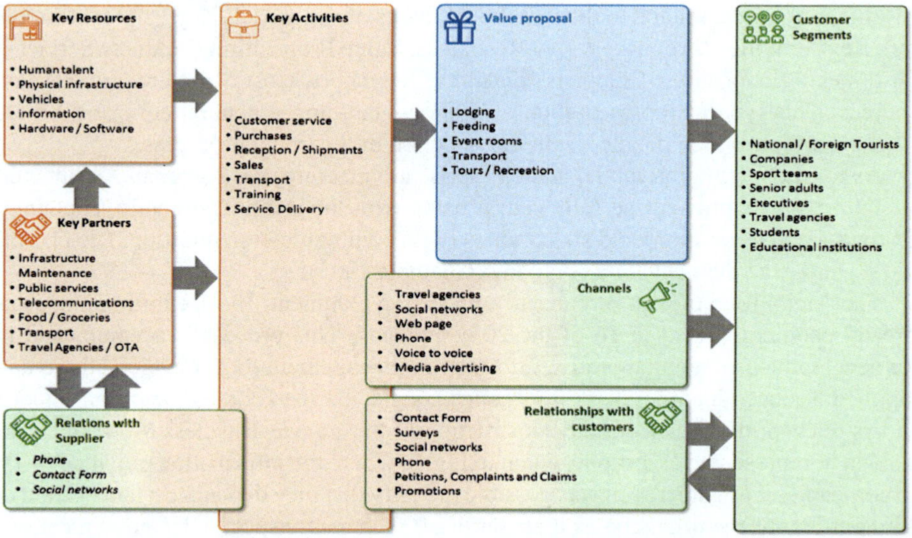

Fig. 2. Canvas business model with pilot sample results.

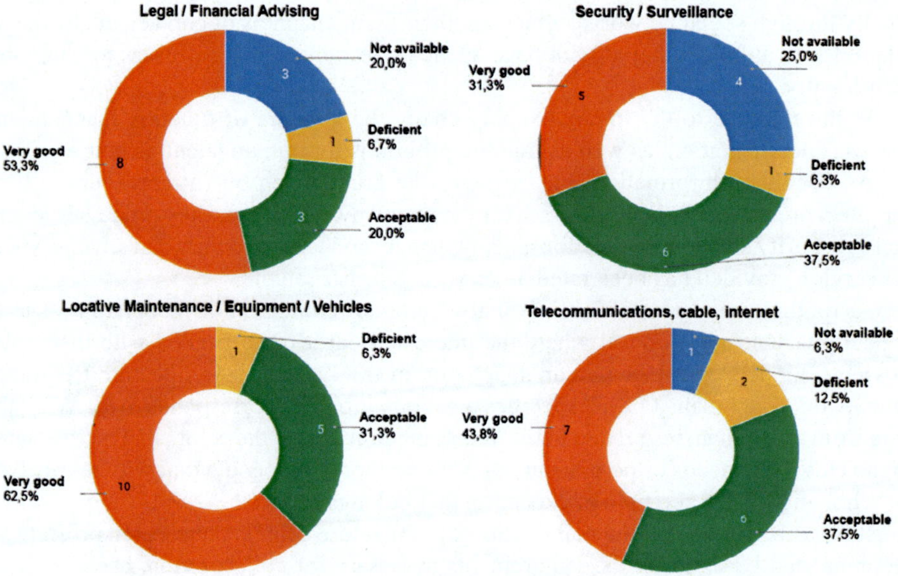

Fig. 3. Grades of service quality to key Suppliers/Partners of the pilot companies.

4 Discussion

Based on the information from the pilot companies collected so far, it has been possible to identify a high dependence on outsourced services in the hotel and tourism sector. It is necessary for companies belonging to it to manage their supply chain efficiently.

Model Architecture for University-Business Collaboration

With a focus on Research Projects

	Stakeholders		Views	Models	Stages	Why Motivation	What Data	How Function	Where Network	Who People	When Time
University	**University** (*General Assembly of Members / Superior Council*)	Planning	Contextual	Scope	Identification	Mission and vision	2028 and 2030 Development Plans	Teaching / Research / Social Projection	Pance Campus, Valle del Lili Campus and San Fernando	Academic Vice Rectory / Administrative Vice Rectory	Project Execution / Publications
University	**Academic Vice Rectory** (*Faculty / Departments*)	Owner	Conceptual	Deal	Definition	Institutional and Program Accreditation / Group Category	Project calls / Congresses / Needs of the region	Research Project / Research Internships / Degree Projects	Valle del Lili Campus / San Fernando / Productive and Services Sector	Teachers, undergraduate and graduate students	Academic calendar
University	**Research Projects** (*Research Group / Programs*)	Designer	Logical	Systems	Representation	Criteria for categorization of the research group	Business data / process metrics / findings, results	Spreadsheets / Collaborative Platform	Network Infrastructure / Cloud Computing	Researchers / Business Team / Technical Support	Project development schedule
University	**Administrative and Financial Vice Rectory** (*Technology planning and development*)	Builder	Physical	Technological	Specifications	Institutional Strategy Directorate	Online database (Cloud Computing)	Physical Infrastructure (Campus) and Technology	Internet / Intranet / Email	Technology planning and development / Technological Innovation Center	Validity of contracts and licensing
Business	**Suppliers**	Integrator	Operational	Components	Configuration	Agreement between University - Google G-Suite	Google Forms, Sheets, Sites, Docs, Drive, Poly, Gmail	Google G-Suite / Add-Ons platform	Google Sites	Google G-Suite / University-Business Platform	Agreement between University - Google G-Suite
Business	**Companies / Society ***	User	Implementation	Operational	Execution	Competitiveness / Cost Reduction / Employment Generation	Process Metrics / KPI's	MejorTour Model / DMAIC Methodology	University-Company	University Work Team - Companies Sector Hotel and Tourism	Project development schedule - Sustainable continuity

*Zachman Architectural Framework Adaptation | * Specific for the MejorTour Project*

Fig. 4. Architecture of the university-company collaboration model.

It is essential to model processes in greater detail, using more structured notations such as IDEF0 (Integration Definition for Function Modeling) and BPMN to identify the reasons for this dependence in detail.

It has been necessary to propose an updated Lean Six Sigma model that incorporates new and modern industrial engineering tools that allow for a broader vision of inter and intra-company integrations, and even the consideration of the concept of Enterprise Architecture that extends the vision to different interested parties.

References

Colombia, D.A.: Ley 1933. Congreso de la Republica de Colombia, Bogotá D.C. (2018)

Fonseca, V.G. A., Correa V.M., Aragón Ch.A.: Adaptación de herramientas Lean Six Sigma en PyME del sector cuero en Colombia. En La vinculación Universidad - Empresa para el desarrollo integral con impacto social, pp. 304–324. Red Universidad-Empresa América Latina y El Caribe - Unión Europea. Obtenido de, México D.F. (2018). http://181.65.155.170/bitstream/UNACH/81/1/Libro_contribución.pdf

Goethals, F.: An overview of enterprise architecture framework deliverables. Obtenido de, May 2003. www.econ.kuleuven.ac.be/leerstoel/sap/FramesPage.htm

Jiju Antony, S.V.: Lean Six Sigma for Small and Medium Sized Enterprises A practical guide, vol. 22. CRC Press, Boca Raton (2016)

Ohno, T.: Toyota Production System: Beyond Large-Scale Production. Productivity Press, New York (1988)

Osterwalder, A., Pigneur, Y., Tucci, C.: Clarifying business models: origins, present and future of the concept. Commun. Assoc. Inf. Sci. **16**, 1–25 (2005)

Pérez-López, E., García-Cerdas, M.: Implementación de la metodología DMAIC-Seis Sigma en el envasado de licores en Fanal. Tecnol. Marcha **27**(3), 88–106 (2014)

Pramod Kumar, S.: Implementing the lean six sigma as a strategy in a small medium enterprises (SMEs). Int. J. Bus. Adm. Manag. **7**(1), 131–140 (2017)

Pyzdek, T., Keller, P.: The Six Sigma Handbook. A Comple Guide for Green Belts, Black Belts and Managers at all Levels. McGraw-Hill, New York (2010)

Robinson, M., Lock, S.: Arts marketing association - AMA culture hive (2014)

Obtenido de. http://culturehive.co.uk/wp-content/uploads/2016/01/Introducing-the-Business-Model-Canvas.pdf

Sarria, M.P., Fonseca, G.A., Bocanegra, C.: Modelo metodológico de implementación de lean manufacturing. Rev. EAN (83), 51–71 (2017)

Urbaczewski, L., Mrdalj, S.: A comparison of enterprise architecture frameworks. Issues Inf. Syst. **7**, 18–23 (2006)

Voehl, F.H.: The Lean Six Sigma Black Belt Handbook: Tools and Methods for Process Acceleration. Productivity Press, New York (2013)

Zachman, J.: Zachman International Enterprise Architecture (2008). https://www.zachman.com/about-the-zachman-framework. Accessed 21 Feb 2019

Green Supply Chain Network Design for the Final Disposal of Used Tires that Mitigate the Environmental Impact in Barranquilla City

Luis Ramirez Polo(✉) and Danna Betancourt M.

Faculty of Industrial Engineering School, Autónoma del Caribe University, Barranquilla, Colombia
{luis.ramirez,danna.betancourt}@uac.edu.co

Abstract. One of the main bases that protect green logistics is the opportunity to continue producing through conservation, which translates as the generation of self-sufficient. The main objective is to reduce the environmental impact in transport activities, and a different approach to the life cycle of the product. This paper show the evaluation of final disposal strategies for tires used in a sector of Barranquilla, Colombia, performing an analysis of the current state and seeking to reduce the environmental impact through techniques such as recycling and reuse, designing a green supply chain by means of multi-objective mathematical models that optimize the profitability and cost of network design and CO_2 environmental emission.

Keywords: Green logistics · Supply chain design · Final disposal strategies · Used tires

1 Introduction

Used tires discarded is a problem of high importance globally. Over the course of time, the commercialization of tires has increased in proportion to the car fleet, which has led to an increasing number of environmental problems, both for the health of the population and for the quality of the air and mantles phreatic. In addition, the inconsistencies at the regulatory level and the lack of management by the environmental authorities have prevented progress in the activities of adequate disposal of this waste considered non-hazardous within Colombian laws. The few attempts at solutions that have been generated in Colombia respect to the uses that can be given to tires after the end of their life cycle are not enough for a country that doesn't have a logistic structure for the collection defined of this special solid waste and actors that propitiate it. In the progress of this article, was important reduce the environmental impact affected by the CO2 emissions, generated by the combustion process in transportation and, in the second instance, to propose an ecological solution for the final disposal of used tires. To this end, a green supply chain network was designed, consisting of a location model and a routing model under the development of mathematical modeling with a multi-objective approach, Defined by (Marler and Arora, 2004) as the process of optimizing

M. T. Castañeda Galvis et al. (Eds.): ICIE 2019, LNMUINEN, pp. 39–46, 2021.
https://doi.org/10.1007/978-3-030-49370-7_5

systematically and simultaneously a collection of objective functions optimization. The importance of adding the green component to the design of the supply chain implies taking environmental management into account in all chain links. (Lin et al. 2013) said that green design is closely related to environmental risk management, product safety, pollution prevention, resource conservation, and waste management; Therefore, through the AHP (Analytic Hierarchy Process) methodology, treatments such as pyrolysis, thermolysis, gasification and mechanical crushing of tires were evaluated, and among them, the most beneficial final disposal method that complied with the policy of not being a contributor of CO_2 emissions was chosen, allowing efficient management of used tires as a product and by-product and preventing pollution.

2 Methodology

Through a survey of 91 establishments found by sample size in a sector of the City of Barranquilla, a diagnosis of the current status was obtained and figures of the number of used tires generated were identified, this being the main input of the location and routing mathematicians' models.

The structure of the methodology shows the scientific relevance focused on the subject, the findings by researchers and the complement with the work presented in this article carried out in Barranquilla City. The results of each model, the alternative selected through the AHP method and a synthesis of the cost-benefit analysis are shown. In this way, compliance with the objective of minimizing CO2 emissions to mitigate the environmental impact is evidenced.

3 Results

Scientific Relevance

The green supply chain management (GSCM) emerged when researchers started to integrate the environmental concerns into the supply chain. The integration of economic, environmental and social facets have been at the forethought of sustainability in today's operations management due to the successful implementation of the green supply chain which eventually has led to overall sustainability of firms (Shamimul et al. 2017). Due to this, organizations have been forced to use practices such as the GSCM (Srivastava 2007) defines them as the Integrating environmental thinking into supply-chain management, including product design, material sourcing and selection, manufacturing processes, delivery of the final product to the consumers as well as end-of-life management of the product after its useful life. (Fleischmann et al. 2010) indicate that what is sought with the management of the green supply chain is to select the best activities for the coordination and design of the supply chain over time, thus being crucial support for decision-making.

(Chin et al. 2015) consider that the GSCM practices are conceptualized to include green procurement, green manufacturing, green distribution, and green logistics. This last practice has been of great relevance for research that seeks to reduce the environmental impact of transport logistics through an optimal network configuration involving the optimization of routing models. (Elhedhli and Merrick, 2012) demonstrated the impact of transport logistics and the incidence of costs on it. The authors designed a green supply chain to reduce carbon emissions, where was confirmed that the addition of carbon costs in the design decision process for the supply chain can change the optimal network configuration, in addition to the appearance of new distribution centers for this improved way the modified distances generate a positive impact that will offset the emissions produced by long distances. In the same way, (Ahmed and Sarkar, 2018) in their mathematical model for the supply chain of second-generation biofuels showed that approximately 88.50% of the network's emissions were caused by transport, while biofuel production contributed to the 7.90% of the carbon in the environment, collaborating with 0.18% of the total cost.

Recognizing the importance of the impact of transport, (Farkavcova et al. 2017) carry out their study highlighting the need to improve the performance of transport systems, indicating that this mainly causes the consumption of fossil fuel resources, reduction potential of the stratospheric ozone layer and impact for non-renewable primary energy used for energy purposes such as natural gas, crude oil, lignite, hard coal, uranium. According to the authors, transport causes significant environmental impacts, in which it must be considered to reduce distances through changing geographically closer suppliers, or backhaul to achieve more effective transport, share the profits from the reduction of transport costs, make investment in technology through a modern fleet and employ the use of biofuels.

Due to the environmental consequences produced by the nature of the transport, a location model and a routing model were developed that allows finding the ideal location of the warehouse to avoid the respective treatment of the used tires and find the best routes to carry out the collection of used. The input data of both models provide the application of surveys to the 91 establishments made up of auto services, gas stations with auto services and automotive workshops. These data were subjected an analysis of Cronbach's Alpha, a statistic that provides a measure of internal consistency of a test or scale and indicates the degree to which all elements measure the same concept, and are therefore interrelated (Tavakol and Dennick, 2011).

Location Model

The location model was configured based on the implemented in the study of the authors (Wang et al. 2011) and corresponds to a mixed linear programming model. In the model, three warehouses were preselected in which the tire treatment center could be located according to the legislation (Decree 0212 of 2014 - POT Barranquilla). The carbon dioxide emission factor obtained from the Mobile6 software, an emission modeling software created by the Environmental Protection Agency (EPA) of the United States, was also considered. This emission factor changes according to the selected vehicle, that in this case, the value was 0.278374 kg/Km. In this study, a 2014 Chevrolet NPR truck type van was used. On the other hand, the value of the cost per mileage was estimated according to an approximation made in a simulation tool called SICE TAC 2.0 of the Transportation Ministry. To Solve the modeling, the Cplex solver of the Gams optimization software will be selected, whose definition is shown below:

Indices

i *Collection points* $i \in D$; j *Treatment center* $j \in A$

Parameters

$S(i)$	*Number of tire collected on i*	$*\,Ct$	*Transportation cost per km*
$FA(j)$	*Fixed opening cost of j*	$*\,Q(j)$	*Capacity of j*
Em	*Emission factor of* CO_2 $(\frac{kg}{km})$	$*\,D(i,j)$	*Distance between i, j*
N_{max}	*Maximun of treatment center*	$*\,DI$	*Rim dimensions*

Decision variables

$X(i,j)$ *Quantities sent to collection point i to treatment center j*
$Y(j)$ *Opening decision of the treatment center j*

Objective Function

$$F1 = \sum_{j=1}^{n} FA(j)\,Y(j) \qquad (1)$$

$$*\,F2 = \sum_{i=1}^{n}\sum_{j=1}^{n} D(i,j) * Ct \qquad (2)$$

$$F3 = \sum_{i=1}^{n}\sum_{j=1}^{n} D(i,j) * Em \qquad (3)$$

$$*\,\min F = F1 + F2 + F3 \qquad (*)$$

Constraints

$$\sum_{j=1}^{n} X(i,j) = S(i), \quad \forall i \in D \qquad (4)$$

$$\sum_{i=1}^{n} X(i,j) * DI = Q(j) * Y(j), \forall j \in A \qquad (5)$$

$$\sum_{j=1}^{n} Y(j) = N_{max} \qquad (6)$$

$$X(i,j) > 0 \qquad (7)$$

$$Y(j) > 0 \qquad (8)$$

The integral multi-objective minimization function (*) consists of three parts: opening costs (1), transport costs (2) and minimization of carbon dioxide emissions (3). On the other hand, the restrictions included: flow (4), capacity (5), number of facilities (6) and those corresponding to non-negativity restrictions (7) and (8). The resulting warehouse was strategically located within the points where the objective function was minimized and in turn input for the routing model. Inside the three options, the selected facility was the most economical in regard especially to the opening costs

Routing Model

The formulation corresponds to a Green Vehicle Routing Problem (GVRP), as a variant of the Vehicle Routing Problem (VRP) and is by the objective of harmonizing the environmental and economic costs by implementing effective routes to meet the environmental concerns and financial indexes (Lin et al. 2013). Based on the GVRP, the basic problem of the VRP is shown as an extension of the m-TSP in which each client i ∈ V\{0} has an associated demand di and each vehicle has a capacity C - homogeneous fleet- (Olivero et al. 2004).

In the formulation of the model, routing was carried out for a single vehicle that will carry out *m* routes with the purpose of contrasting the reduction of distances with the objective of minimizing CO_2 emissions. The definition of the mixed- integer linear programming model is of the deterministic type, since the weight QVi, j, k of the tires transported by the vehicle k towards i, j is known. The base model was proposed by the authors (Kopfer and Kopfer, 2013) and is shown below:

Indices

i, j *Start and end notes* $i, j \in N\{0\}$; k *Vehicles* $k \in V$

Parameters

$* Dij$ *Distance between i y j* $* Ek$ *Fuel consumption of empty Vehicle k*
$* Qj$ *Cantidad de llantas a recoger en j*
$* Ck$ *Fuel consumption of the loaded vehicle k*
$* QKk$ Maximun load capacity of k $* n$ *Unknown Value*

Decision variables

$* QVi, j, k$ *Cargo of vehicle k traveling between i, j* $* Ui$ *Arbitrary real variable*
$* Xi, j, k$ *1 if the vehicle k serves node j inmediatly after serving i*
$* Sj, k$ *1 if node j is served by vehicle k*

Objective Function

$$\sum_{i=0}^{n}\sum_{j=0}^{n}\sum_{k=1}^{m} Dij * (Ek * Xijk + Ck * QVijk) \quad (1)$$

Constraints

$$\sum_{j=1}^{n} Qj * Sjk \leq QKk \quad (2) \qquad \sum_{i=0}^{n} Xijk = \sum_{i=0}^{n} Xjik \quad (3)$$

$$QVijk - QKk * Xijk \leq 0 \quad (4) \qquad \sum_{i=0}^{n} Xijk = Sjk \quad (5)$$

$$\sum_{k=1}^{m} Sjk = 1, \forall j \in N \backslash \{0\} \quad (6) \qquad \sum_{j=1}^{n} X0jk \leq 1 \quad (7)$$

$$\sum_{i=0}^{n} QVijk - \sum_{i=1}^{n} QVjik = Qj * Sjk, \quad \forall k \in V; \forall j \in N \backslash \{0\} \quad (8)$$

$$Ui - Uj + n \sum_{k=1}^{m} Xijk \leq n - 1, \quad \forall i \in; \forall j \in \{0\} \quad (9)$$

$$QVijk \geq 0 \quad (10) \qquad Ui \geq 0 \quad (11)$$

The objective function multiplies the distance by two factors: the first one refers to the fuel consumption of the empty vehicle and the second shows the behavior of the fuel consumption of the loaded vehicle. The restrictions (2), (3) and (4) are of capacity, the constraints (5), (6), (7) and (8) are of balance of flow, the restriction (9) is the elimination of sub-routes that at the same time impose that the total demand of the clients visited by a vehicle cannot exceed the capacity C (Olivero et al. 2004) And the (10) and (11) are non-negativity restrictions. Restrictions (2), (3), (5), (6), (7) and (9) are common in the formulation of a basic VRP problem.

For the computational experiment, the Gurobi solver of Google OR-Tools module of the Python programming language was used, however, to the analysis 91 instances and 1 deposit no response was obtained in computational time with the presented formulation (as it is a model of optimization of medium size), so it was solved using a simple VRP, and then applying the objective function of the GVRP to the routes delivered by the VRP, comparing them with the direction or orientation of the route according also with the vehicle weight that will generate a lower CO2 consumption.

In this way, the results show that the route provided by the simple VRP is not always ideal when confronted with the objectives of minimizing carbon dioxide emissions. For this case, it was necessary to change the direction of two (2) of the six (6) routes delivered by the model, obtaining a reduction of 8.25% in the emission factor by 8.96 km, corresponding to 2.50 kg of CO_2.

The best routes that appear in Table 1, show in total less distance, fuel consumption and therefore a positive impact on the total cost.

Table 1. Routing Model Summary

Model results comparison		
Distance (Km)	108.671	117.635
Cargo (Units)	1137	1137
Liters (Diesel)	14.33	17.80
Total Cost	$ 32,841.21	$ 40,790.13
Emission Factor	30.25	32.75

Source: Self made

Once the collection routes of the tires used in the 91 active points (auto services, gas stations with auto services and automotive workshops) were obtained, an analysis of final disposal treatment strategies of this special solid waste was performed and selected the best alternative through the process of Analytical Hierarchy (AHP), for being an intuitive method to formulate and analyze decisions (Hajeeh and Al-Othman, 2005). The strategies contemplated for evaluation were pyrolysis, thermolysis, gasification and mechanical crushing of tires.

To develop the Analytical Hierarchy process, three factors were contemplated: machinery costs, polluting emissions and economic viability. The weighting was made according to the review in the bibliographic sources and expert opinion.

The way to evaluate the matrices was analyzing the negative impact of objective i in correspondence with objective j according to the weighting interpretation matrix (Winston, 2004). The more positive numbers indicate higher importance with respect to that negative impact. Each of the factors was tested for consistency to validate the random index in accordance with the consistency index.

The final disposal strategy selected was mechanical crushing because it represented less damage in terms of the three factors evaluated, and it presented a better weighting regarding the pollutant emissions factor. According to the authors (Camargo et al. 2017) within the mechanical crushing, no CO2 emissions are mentioned in accordance with the literature review.

In accordance with the guide of used tires of 2006 of the Chamber of Commerce of Bogotá, this process consists of using blades to shred the tires; generally it is done cascading, that is, the tires are crushed gradually until they reach the minimum required size and then pneumatic and magnetic sorters are used to separate the textile.

Temperature crushing is a well-developed technology at the industrial level, so it is quite easy to implement (Tirel, 2017). The phases are: Primary crusher, secondary crusher, granulator, separation. According to (Siddique and Tarun 2004) the tires after their treatment can be classified as:

- Patchwork tires: with the subcategories of torn tires and broken tires (300–460 mm by 100–230 mm wide, down to smaller sizes such as 100–150 mm).
- Rubber screening (19 mm to 0.15 mm).
- Granulated rubber (4.75 mm less than 0.075 mm).

Some of the products from the used tires are asphalt for roads and highways, vibration absorption systems, modification of the properties of the concrete, fillings for artificial turf of soccer fields, athletics or cycling tracks, road speed reducers, solutions for stables, floor tiles, decorative mulch and playgrounds (Eco Green Equipment 2018).

The cost-benefit analysis of the waste tire shredder and the logistic process, shows that, although the costs of machinery investment, vehicle purchase, and lease expenses are significant, the net flow of the project shows profits that grow exponentially provided that the objective of increasing annual sales, and recognizing that inflation is an important protagonist to analyze the behavior of earnings.

4 Discussion

The multiobjective mathematical modeling finds an appropriate place to locate the treatment center of the used tires. The location model obtained the best point and with the routing model, it was optimized in a total number of 6 trips with the use of a single vehicle obtaining a reduction of 8.25% of CO_2 emissions. Finally, it was determined the best way to satisfy the objective of reducing CO_2 emissions is to perform mechanical crushing as a final treatment of used tires, according to the literature that decreases is a zero-emission policy. The cost-benefit analysis also shows a trend with exponential growth under the control of the factors of increased sales and inflation.

This research is expected to be an input for optimization proposals, and solution alternatives that develop more appropriate strategies for this sector that is constantly growing, as well as the migration developed clean technologies such as the use of electric vehicles, to analyze and reduce new minor environmental problems that could be generated. Under the current model, the possibility of expanding the analysis to other environmental impact factors such as the emissions of Hydrocarbons (HC) and Nitrogen Oxides (NOX) contributing to the greenhouse effect, also emitted and generated by the internal combustion of the engines. It is proposed to carry out a similar study for the motorcycle market, since they are one of the main generators of the number of used tires, especially since they are not considered as the tires used by the Ministry of Environment and by of Colombian legislation.

Other contributing studies must be of the stochastic type recognizing that the demand is dynamic, which generates an increasingly real and worthwhile situation of being studied, developed and applied.

References

Ahmed, W., Sarkar, B.: Impact of carbon emissions in a sustainable supply chain management for a second generation biofuel. J. Clean. Prod. **186**, 807–820 (2018)

Camargo, S., Franco, J., Chud, V., Osorio, J.: Modelo de simulación dinámica para evaluar el impacto ambiental de la producción y logística inversa de las llantas. Ingeniería y Desarrollo **35**(2), 357–381 (2017)

Chin, T., Tat, H., Sulaiman, Z.: Green supply chain management environmental collaboration and sustainability performance. Procedia SIRP **26**, 695–699 (2015)

Eco Green Equipment *Eco Green Equipment.*: Retrieved from El gránulo y sus distintas aplicaciones (2018).http://ecogreenequipment.com/es/el-granulo-de-caucho-y-sus-distintas-aplicaciones/

Elhedhli, S., Merrick, R.: Green supply chain network design to reduce carbon emissions. Trans. Res. Part D **17**(5), 370–379 (2012)

Farkavcova, V.G., Rieckhof, R., Guenther, E.: Expanding knowledge on environmental impacts of transport: a Case study using life cycle assessmet. Trans. Res. Part D. **61**, 68–83 (2017)

Fleischmann, B., Meyr, H., Wagner, M.: Supply Chain Management and Advance planning. Springer-Verlag, Berlin Heidelberg, Berlin, Germany (2010)

Hajeeh, M., Al-Othman, A.: Application of the analytical hierarchy process in the selection of desalination plants. Desalination **174**(1), 97–108 (2005)

Kopfer, H., Kopfer, H.: Emissions Minimization Vehicle Routing Problem in Dependence of Different Vehicle Classes. *Dynamics in Logistics*, 49–58. (2013).

Lin, C., Choy, K., Ho, G., Chung, S., Lam, H.: Survey of green vehicle routing problem: past and future trends. Expert Syst. Appl. **41**(4), 1118–1138 (2013)

Lin, R.-J.: Using fuzzy DEMATEL to evaluate the green supply chain management practices. J. Clean. Prod. **40**, 32–39 (2013)

Marler, R., Arora, J.: Survey of multi-objective optimization methods for engineering. Struct. Multidisc Optim. **26**(6), 369–395 (2004)

Olivero, A.: Heurísticas para Problemas de Ruteo de Vehículos. Montevideo, Uruguay (2004).

Shamimul, I., Noorliza, K., Firdaus Bin Ahmad, F., Mohamed, S.: A review on green supply chain aspects and practices. De Gruyter Open. **12**(1), 13–36 (2017)

Siddique, R., Tarun, R.N.: Properties of concrete containing scrapt-tire rubber an overview. W. Manag. **24**(6), 563–569 (2004)

Srivastava, S.K.: Green Supply Chain Management: a state of art literature review. Int. J. Manag. Rev. **9**(1), 53–80 (2007)

Tavakol, M., Dennick, R.: Making sense of Cronbach's alpha. Int. J. Med. Educ. **2**, 53–55 (2011)

Tirel, K.: Ingeniería de perfil de modernas plantas para reciclaje de neumáticos fuera de uso (NFU). Santiago de Chile, Chile (2017).

Wang, F., Lai, X., Shi, N.: A multi-objective optimization for green supply chain network. Decisions Support Syst. **51**(2), 262–269 (2011)

Winston, W.: Investigación de operaciones. Aplicaciones y Algoritmos. Thomson International, Connecticut (2004)

Sustainability of an Educational Program Applying Lean Thinking to Its Value Chain

Vera J. Santiago Martinez[1](✉), John A. Zamora Salas[2], and Genett I. Jimenez Delgado[3]

[1] Industrial Engineering Program, Faculty of Engineering, Universidad Autónoma del Caribe, Barranquilla, Colombia
vera.santiago@uac.edu.co

[2] Servicio Nacional de Aprendizaje SENA, Barranquilla, Colombia
jzamora@sena.edu.co

[3] Industrial Process Engineering, Faculty of Engineering, Institución Universitaria ITSA, Barranquilla, Colombia
gjimenez@itsa.edu.co

Abstract. The main topic of this research is lean thinking for sustainability and continuous improvement of a professional education program in a Colombian institution. The research objective was to replicate the models found on literature and to develop new one. The models are based on experiences of universities outside country. The contribution of this research was an innovation on the management of the institution.

Keywords: Lean thinking · Higher education · Sustainability · Value chain · Continuous improvement

1 Introduction

Most institutions of higher education in the country have implemented a quality management system certified by an entity endorsed for it, such as ISO 9001, and models of accreditation monitored by the Ministry of National Education. The implementation of these quality models has led institutions to carry out exhaustive self-evaluation processes and audits, in which multiple opportunities for improvement in the processes, products and services of professional training have been identified, opportunities that become the point starting for a number of actions that lead to excellence (Jiménez 2014).

In addition, according to Jimenez and Zapata (2016) the use of integrated methodologies for continuous improvement can offer organizations different approaches for diagnosis, analysis of root-causes and propose improvement scenarios such as Balanced Scorecard, Integrated Management Systems (Jiménez et al. 2018), the use data analysis techniques (Jimenez et al. 2018) the application of Multicriteria Decision-Making Methods – MCDM (Ortiz et al. 2017), (Jiménez et al. 2019), and the use of lean think model. The lean thinking model is geared towards achieving high quality standards for higher education institutions.

M. T. Castañeda Galvis et al. (Eds.): ICIE 2019, LNMUINEN, pp. 47–56, 2021.
https://doi.org/10.1007/978-3-030-49370-7_6

In the world, various types of organizations have implemented Lean-thinking initiatives (Duro and Gilart 2016). Waterbury (2015) states that higher education institutions have been experimenting with Lean since the mid-2000s. Universities, like the rest of the organizations, whether public or private, for profit or not, have great difficulties to incorporate cycles of continuous improvement to the processes associated with decision making and visualize the progress towards their social ends (Araujo et al. 2011).

Lean thinking offers a methodology that favors continuous improvement and thus its sustainability. Comm and Mathaisel (2005) formulate that services and costs within higher education fall within the scope of lean sustainability. The same lean sustainability concepts applicable to the business sector are transferable to the education sector. The benefits of Lean for institutions that offer higher education can be quantified significantly in a reduction of waste by up to 50% and errors between 60 and 90% which cause the increase in service costs of between 30 and 80% (Carrillo et al. 2010). This means that educational institutions manage to provide improved services with lower costs, higher operating margins. In this sense, lean sustainability is a good option for higher education, and this helps universities to identify which areas contribute to the institution's prosperity and survival, and whether or not to continue investing in these areas is profitable.

2 Methodology

2.1 Literature Review

According to Santos et al. (2014), the philosophy for reducing waste and making the most efficient production better known as Just in Time or Toyota Production System was created by Taiichi Ohno and Shigeo Shingo in the 50s, but This has its beginning in the production flow systems of Henry Ford and the writing on scientific management of Fredrick Taylor, since Mr. Taiichi recognized that the stimulus of his ideas was in the thorough reading of Ford's ideas. The term Lean Manufacturing was accepted in the United States, applied to a productive system means "agile, flexible", that is, capable of adapting to customer needs. In 1996, the concept of Lean Thinking or Lean Thinking was proposed as a powerful tool for creating value, while eliminating everything that could be superfluous in an organization, Womack and Jones (2003) on the other hand Carrillo et al. (2010) establish that it is a management methodology to evolve in the implementation by stages, from a traditional execution of processes to a managed according to lean principles, with highly efficient and competitive processes, where the connection between the removal of molt and the creation of value. In 1997, the first writings on the Lean Services (Lean Service) emerged and between 2010 and 2013, he achieved his greatest interest. According to Hadid and Mansouri (2014), it is a sociotechnical system that is divided into 2 distinct groups, practices related to the human field such as education, training, empowerment and multifunctional employees, and technical practices. In 2000, Dahlgaard and Ostergaard (2000) related the organization in higher education with seven types of waste and five Lean principles. Establishing that when reviewing the lean methodology, educational institutions can be considered as good candidates for the application of Lean, and that this would lead to a significant reduction of waste and improvements in terms of quality of services.

In 2012 Womack and Jones, in their book establish the term Lean Thinking or (Lean Thought), as the most powerful tool available to us to create value, while eliminating everything that can be superfluous in an organization. With respect to higher education, Waterbury (2015), states that higher education institutions have been experimenting with Lean since the mid-2000s, schools from several countries experimented with Lean in higher education even before there was an educational roadmap, The importance of models that provide basic principles (value, value flow, flow, traction and perfection) that guided waste disposal is highlighted. In the world various institutions have implemented Lean Initiatives, Sunder (2016) when studying Antony (2014), identified some universities that include the University of St. Andrews (Scotland), the University of Cardiff (Wales), the University from Coventry, the University of Portsmouth (England), the Central Connecticut State University, Bowling Green State University, MIT and Oklahoma State University (USA) also identified some reports of the insipient phases of lean implementation on the websites of Kings College, London, Miami University, USA, National University of Singapore, Valdosta State University, Georgia, Heriot Watt University, United Kingdom, Gordon State College, USA, Sri Sathya Sai Institute of Higher Education, India, additionally Duro and Gilart (2016) states that according to (Hines and Lethbridge 2008), there have been few examples of a true Lean system in academic organizations.

In this regard, it is highlights several universities such as the University of Wisconsin-Madison (USA), Shanghai JiaoTong University, Lally School of Management and Technology (Connecticut Rensselaer Polytechnic Institute, USA), University of Iowa, University of New Orleans, and the University of Scranton (Pennsylvania), however, no literature is found in books, magazine articles or research theses on the implementation of Lean in Higher Education Institutions when researching databases such as ProQuest, E- book and the Google search engine, no Lean implementation initiative was identified in the education sector in Colombia.

The benefits of Lean in institutions that offer higher education can be significantly quantified as expressed by Carrillo et al. (2010) and George (2003), who argue that in any HEI it can be reduced to waste by 50%, and sometimes much more, this according to the problems found in the processes and according to the tools read that are used for improvement, as long as they continue to apply and maintain the Lean principles in the processes, on the other hand, George (2003) speculated that if Lean were introduced in higher education, the errors could potentially be reduced by an astonishing 60–90%, he also claimed that waste increased the cost of services by 30–80%. Dragomir and Surugiu (2013) raise a specific case of analysis of the implementation of lean in higher education at the University of Central Oklahoma, this institution has adopted the concept of Lean Thinking and was transformed into Lean University due to several reasons: budget reduction, insufficient financing to cover mandatory cost increases, the outdated administrative process, job dissatisfaction of employees and low levels of productivity. Lean Thinking was initially implemented focusing on the administration processes but was later expanded to other divisions.

2.2 Research Methodology

The type of research used, according to the purpose and scope, is applied, since the proposals made in this work may be applied to Higher Education Institutions. Given the level of knowledge, research is descriptive, since it seeks to diagnose the state of Lean thinking in higher education.

According to the design, the research is mixed, since it has both quantitative and qualitative components. Various strategies such as documentary and bibliographic, field review, survey and interview are used as information gathering techniques.

The research phases follow the guidelines of the four-step model for the implementation of lean in higher education at the Central University of Oklahoma. In step one, the opportunities were identified by completing a diagnostic search of the entire organization to find problems and opportunities, then in step two a solution was designed, creating a draft for success, that involves all employees with training processes, mapping and planning. In step three the Kaizen events, central and metric teams were implemented to implement and illustrate the change and finally in step four, it is continuously improved through performance controls after the projects are completed. Following the model described, the four stages were performed: diagnosis, identification of solutions, implementation and continuous improvement, but making adjustments that were identified within the context of the institution and the case of the educational program. In lean thinking it involves a practical work applying the methodology from which the conceptual basis was subtracted to formulate the implementation model in said institution.

3 Results

At this point, the characterization of the training service and the map of the value chain of a program in Aeronautics of a Higher Education Institution was prepared, laying the foundations for the construction of the matrix cause wastage effect, which in turn It served as the main input to propose simple, innovative and creative improvement actions, applying lean thinking and tools, to eliminate waste and mitigate the causes to the problems identified in the current value chain. The proposals for each prioritized waste will be presented, these proposals include actions such as: the foundations of the strategic and prospective plan of the program, an analysis of the technological infrastructure, a proposal of balancing and cost of the offer, as well as technological solutions for the management and analysis of program information, in order to propose a future value chain without waste in the next section of this research.

The proposal for the characterization of the value chain of the Vocational Training Service is presented as a result of the analysis in the worktables with administrative and instructors and the documentary review of the process, procedure and document files related to the training program. In this regard, the characterization of the Value Chain of the Vocational Training Service of the aircraft online technologist program is shown in Fig. 1. Five discriminated processes are presented as follows: 1 innovation with two activities, 3 value with five activities and 1 support with three activities, this will be the starting information for the establishment of the map of the current value chain, see Fig. 1, which is the second product that will serve to respond to the objective, and that is composed of a flow of value of the service and the activities and sub-activities that are carried out in each process, being This is the clearest way for the identification and association of waste as presented in the identification matrix waste management and classification.

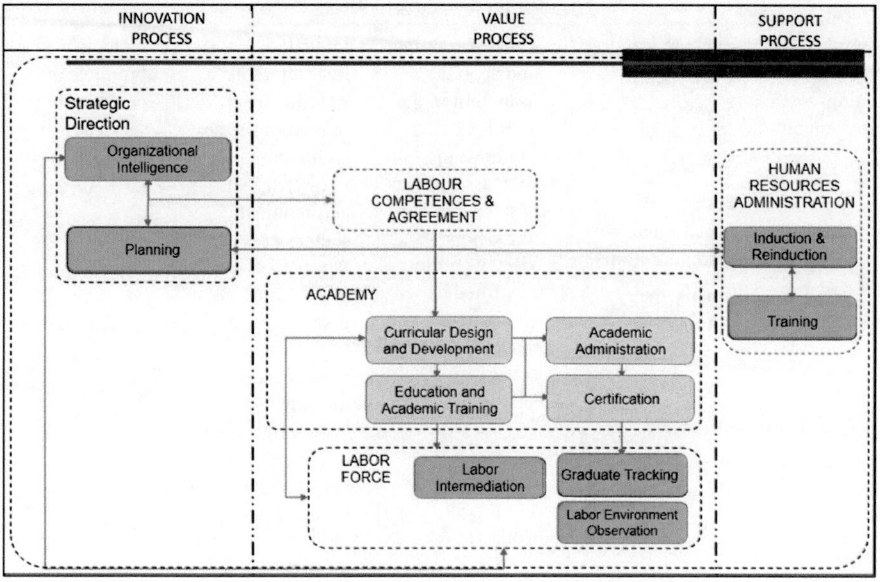

Fig. 1. Characterization of the value chain of the vocational training service. Source: Prepared by authors.

Table 1. Waste identification and classification matrix. Source: Prepared by the authors.

Process	Activity	Critical aspects of the service identified	Waste or lack	Waste classification
Strategic address planning	Perform organizational intelligence Perform the planning	• Few training programs titled and complementary • Excess groups waiting to do their practice	• Catalog of limited training programs • Insufficient obsolete techniques, tools and equipment	• Bad planning in general • Overproduction • Bad knowledge management
Management of concertation and labor competences	Normalize labor competencies	Few competition rules	• Bad occupational characterization • Disarticulated norms of the productive sector	Defective products or poorly provided services

(*continued*)

Table 1. (*continued*)

Process	Activity	Critical aspects of the service identified	Waste or lack	Waste classification
Integral professional training management	Execute training Design and develop curriculum content Perform the educational administration Certify the training	• Lack of training and specific induction in the technical areas of knowledge • Increase the practices in the workshop • They were not identified • Delay in the delivery of the certification	• Outdated instructors in specific areas related to the new technical knowledge of the aeronautical sector • Absence of guides and instruments to execute and evaluate the training • Poorly designed selection instruments • Errors in the registration of the information sent	• Defective products or poorly provided services • Timeouts • Reprocessing
Employment management, occupational analysis and employability	Evaluate the occupational dynamics of the sector	Poor reports and the samples used are not representative of the population of graduates of the program	• Low employment location of graduates • Insufficient occupational studies	Bad knowledge management
Human talent management	Perform induction and re-induction of officials Conduct training for officials	Lack of training and specific induction in the technical areas of knowledge	• Generic and poorly planned induction process • Training plans for disjointed training instructors	Bad knowledge management

Table 1 shows the consolidation of all the improvement opportunities raised by the apprentices, businessmen, administrators and instructors, as well as the improvement opportunities identified, in the conclusions obtained from the analysis of the primary information of the productive sector and internal program information, according to the diagnosis made in the previous chapter, obtaining a total of 32 improvement actions. They were then classified according to the process and activity that they affect according to the current value chain, followed by this, each of these opportunities for improvement identified, they were given a connotation of waste, considering that these aspects do not add value to the service training and actions must be established to eliminate them. Finally, a classification of waste types was made.

In Table 2 shows the action to mitigate waste, specifying the lean tool that served as a methodology and inspiration for the solutions:

Table 2. Waste matrix and mitigation actions. Source: Prepared by the authors.

Waste or lack	LEAN tool	Proposed actions for improvement
Absence of a strategic program plan	PESTEL DOFA	Bases for the strategic plan of the aerospace network
Dismantling of the training center and the productive sector	KPI VISUAL MANAGEMENT	Technological gap analysis tool development (spreadsheet for analysis). Formulation of the aerospace network technology plan. (Hernández and Jiménez 2017)
Excess of apprentices in training	KANBAN	Proposal for balancing the offer of professional training to meet the real demand of the aerospace business sector
High volumes of information process	BIG DATA	Web application development (quality assurance software)
Excess of graduates in the labor market	KPI	Proposal for balancing the offer of professional training to meet the real demand of the aerospace business sector
Outdated instructors in specific areas related to the new technical knowledge of the aeronautical sector	DOFA	Proposed training plan instructors verified aviation program with aeronautical licenses
Absence of detailed occupational studies	PESTAL DOFA	Bases for the strategic plan of the aerospace network
About executing the training offer	KANBAN	Program analysis tool development (Excel spreadsheet with analysis model)
Ignorance of the cost of training	KPI	Development of training cost analysis tool (Excel spreadsheet with analysis model)

In Fig. 2 shows the new chain value after applying lean tools:

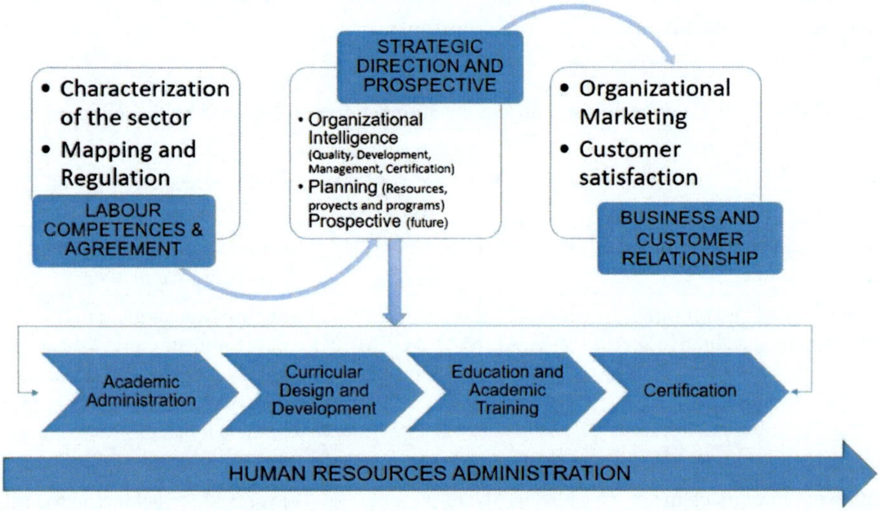

Fig. 2. Future value chain of the map of the vocational training service aircraft online technologist program. Source: Prepared by authors

4 Discussion and Conclusions

Lean thinking seeks to create value for customers with fewer resources by minimizing waste. A more efficient process not only makes an organization more efficient but also contributes to its sustainability and that of its operations. This same principle applies both to the context of manufacturing and services, and as demonstrated in the development and results of research in the field of educational services. A Lean process relies on involving the relevant entities or parts of the processes and a repetitive work to identify problems and propose solutions. At the institutional level it implies a provision of information and resources that favor the documentary analysis and the mapping of the service as a fundamental input for the development of a future value chain. The solutions with which they faced the prioritized waste are associated with transversal problems in the educational and training services offered by the institution which favors that an implementation model based on the experience with the studied program can be favorably replicated in other academic programs of the same that is proposed for the current one. Elements such as the time of execution, implementation and completion may differ but the inputs constructed as the analysis tool, the cost tool, the methodology of curricular designs, the tool of technological gaps and the kamban model for supply planning are also applicable and constitute the work methodology developed for the institution.

The determination and prioritization of waste, gave rise to 9 improvement opportunities, which are described below: 1 a strategic plan of the Aerospace Network, 2 a tool for analysis of Technological Gaps, 3 a Technology Plan of the Aerospace Network associated with strategic plan, 4 a proposal for balancing the offer of professional training to respond to the real demand of the aerospace business sector, 5 a web Application (Quality Assurance Software), 6 a Program Analysis Tool (Excel Spreadsheet with analysis model), 7 a Training cost analysis tool (Excel spreadsheet with cost analysis), 8 a proposal for new curricular designs to meet the new needs of the sector and finally 9. A training proposal for instructors With these solutions the prioritized waste was confronted, however it cannot be affirmed with certainty that they have given solution to the problems, because the solutions have long periods of implementation because they are associated to solve problems of the training service, which has a long period of time in its execution and completion, however they have been well received by senior management and promise to have a favorable period of implementation and adjustments to the extent of their execution.

References

Antony, J.: Readiness factors for the lean six sigma journey in the higher education sector. Int. J. Prod. Perform. Manag. (2014). https://doi.org/10.1108/ijppm-04-2013-0077

Araujo, P.: Universidades Lean: Contribución para la reflexión. Revista de la Educación Superior, vol. XL 4 (2011)

Carrillo, M.S., Pons, R., Barrios, P.A., Puello, A.P.: Lean thinking metodología de gestión de mejoramiento en instituciones de educación superior. Engineering and Technology (2010)

Comm, C.L., Mathaisel, D.: An exploratory study of best lean sustainability practices in higher education. J. Qual. Assur. Edu. (2005). http://doi.org/fvktrv

Dahlgaard, J., Ostergaard, P.: TQM and lean thinking in higher education. In: The Best on Quality: Targets, Improvements, Systems, vol. 11. American Society for Quality (2000)

Dragomir, C., Surugiu, F.: Implementing Lean in a Higher Education University. Constanta Maritime University's Annals. Researchgate (2013). https://www.researchgate.net/publication/249007449_IMPLEMENTING_LEAN_IN_A_HIGHER_EDUCATION_UNIVERSITY

Duro, V., Gilart, V.: La competitividad en las instituciones de educación superior. Aplicación de filosofías de gestión empresarial: lean, Six Sigma y Business Process Management (BPM). Revista Economía y Desarrollo, vol. 157, no. 2 (2016)

George, M.L.: Lean Six Sigma for Service: How to Use Lean Speed and Six Sigma Quality to Improve Services and Transactions. McGraw-Hill, New York (2003)

Hadid, W., Mansouri, S.A.: The lean-performance relationship in services: a theoretical model. Int. J. Oper. Prod. Manag. (2014). https://doi.org/10.1108/IJOPM-02-2013-0080

Hernandez, L., Jimenez, G.: Characterization of the current conditions of the ITSA data centers according to standards of the green data centers friendly to the environment. In: Silhavy, R., Senkerik, R., Kominkova Oplatkova, Z., Prokopova, Z., Silhavy, P. (eds.) Cybernetics and Mathematics Applications in Intelligent Systems. CSOC 2017. Advances in Intelligent Systems and Computing, vol. 574. Springer, Cham (2017). https://doi.org/10.1007/978-3-319-57264-2_34

Hines, P., Lethbridge, S.: New Development: Creating a Lean University. Public Money and Management. Researchgate (2008). https://www.researchgate.net/publication/4779674_New_Development_Creating_a_Lean_University

Jiménez, G.I.: Procedimientos para el mejoramiento de la calidad y la implantación de la Norma ISO 9001 aplicado al proceso de asesoramiento del Centro de Investigaciones y Desarrollo Empresarial y Regional en una Institucion de Educacion Superior basados en la gestion por procesos. In: Congreso de Gestion de la Calidad y Proteccion Ambiental GECPA 2014, Habana, Cuba (2014)

Jiménez, G.I., Zapata, E.: Metodologia integrada para el control estratégico y la mejora continua, basada en el Balanced Scorecard y el Sistema de Gestión de Calidad: Aplicacion en una organizacion de servicios en Colombia. In: 51a Asamblea Anual del Consejo Latinoamericano de Escuelas de Administracion CLADEA 2016, Medellín, Colombia (2016)

Jiménez, G.I., Novoa, L.N., Ramos, L.V., Martínez, J., Alvarino, C.: Diagnosis of initial conditions for the implementation of the integrated management system in the companies of the land cargo transportation in the city of Barranquilla (Colombia). In: Stephanidis, C. (eds.) HCI International 2018 – Posters' Extended Abstracts. HCI 2018. Communications in Computer and Information Science, vol. 852. Springer, Cham (2018). https://doi.org/10.1007/978-3-319-92285-0_39

Jimenez, G.I., Hernandez, L.E., Hernandez, H.G., Cabas, L., Ferreira, J.: Evaluation of quality management for strategic decision making in companies in the plastic sector of the colombian caribbean region using the TQM diagnostic report and data analysis. In: Stephanidis, C. (eds.) HCI International 2018 – Posters' Extended Abstracts. HCI 2018. Communications in Computer and Information Science, vol. 852. Springer, Cham (2018). https://doi.org/10.1007/978-3-030-22216-1_32

Jimenez, G.I., Balmaceda, N., Hernández, H.G., de la Hoz, E., García, J., Martinez, J.: An integrated approach of multiple correspondences analysis (MCA) and fuzzy AHP method for occupational health and safety performance evaluation in the land cargo transportation. In: Duffy, V. (eds.) Digital Human Modeling and Applications in Health, Safety, Ergonomics and Risk Management. Human Body and Motion. HCII 2019. Lecture Notes in Computer Science, vol. 11581. Springer, Cham (2019). https://doi.org/10.1007/978-3-030-22216-1_32

Ortiz, M.A., Kucukaltan, B., Carvajal, D., Neira, D., Jimenez, G.I.: Strategic hybrid approach for selecting suppliers of high-density polyethylene. J. Multi-Criteria Decis. Anal. (2017). https://doi.org/10.1002/mcda.1617

Santos, J., Wysk, R.A., Torres, J.M.: Improving Production with Lean Thinking. Wiley, Hoboken (2014)

Sunder, M.V.: Lean six sigma project management – a Stakeholder management perspective. TQM J. (2016). https://doi.org/10.1108/tqm-09-2014-0070

Waterbury, T.: Learning from the pioneers. Int. J. Qual. Reliab. Manag. (2015). http://doi.org/czzh

Womack, J.P., Jones, D.T.: Lean Thinking. Free Press, New York (2003)

Sustainability

Characterization of Composite Material Reinforced with Fique Fibra Through the Winding Filament Manufacturing Technique

Sergio A. Gómez[1(✉)], Roger Peña[2], Alfonso Santos[1], and Edwin Córdoba[1]

[1] Faculty of Mechanical Engineering, School of Engineering, Pontifical Bolivarian University – Sectional Bucaramanga, Floridablanca, Colombia
{sergio.gomezs,alfonso.santos,edwin.cordoba}@upb.edu.co
[2] Technology in Industrial Production, School of Engineering, Technological Units of Santander, Bucaramanga, Colombia
industrial@correo.uts.edu.co

Abstract. Recently composite materials reinforced with natural fibers have taken great attention due to their technical, economic characteristics and low impact on the environment, becoming attractive elements to be used in industrial applications. One of the fibers used for this purpose has been the fique fiber, making different laminated composites, however, they are not reported in the literature manufacturing tubular parts with this fiber type. For this reason, in this investigation was manufactured a tubular piece with bio-composite material reinforced with fique fiber using filament wilding technique, characterizing mechanically and morphologically to know its behavior which allows determine future possible applications. The material was manufactured with continuous fique fibers and a polyester resin; morphology was known through scanning electron microscopy and mechanical behavior was evaluated by compression compared to conventional PVC tube in a universal machine. The results showed that the biocomposite exhibits lower mechanical behavior compared to PVC and poor adhesion between the matrix and the fiber.

Keywords: Composite · Fique · Filament wilding

1 Introduction

Fique is one of the most representative fibers in tropical America, especially in Colombia, commonly used for ropes, handicrafts and packaging manufacture for the coffee and cocoa transport (Gañán and Mondragon 2004). This fiber has an approximately density of 0.870 g/cm^3, a tensile strength of 237 MPa and a elasticity module of 8,010 GPa (Muñoz et al. 2014).

Due to its overproduction and synthetic fiber as a substitute product, new applications are being sought where it can be used industrially, one of them being as a reinforcement in composite materials (Hidalgo-Salazar and Correa 2018).

M. T. Castañeda Galvis et al. (Eds.): ICIE 2019, LNMUINEN, pp. 59–65, 2021.
https://doi.org/10.1007/978-3-030-49370-7_7

A composite consists in a mixture of two or more materials (normally fiber and matrix) to obtain a new material with improved properties regarding the individual behavior of each of them (Sapuan 2017).

Various manufacturing processes are currently used for the composite materials manufacture reinforced with natural fibers, being the most representative: pultrusion, vacuum infusion, hand lay-up, molding compression, injection and filament winding (Summerscales and Grove 2014).

The last manufacturing process (filament winding) consists pre-impregnated winding continuous fibers with a polymer matrix into a rotating chuck with the aim of creating a solid structure (Dun *et al.* 2019). The properties of the final manufactured product will depend on the winding angle and the fiber tension. Its main application is the pressure tanks and tubular parts manufacture (Sorrentino *et al.* 2019).

Composite materials reinforced with fique fibers have typically been manufactured with compression molding techniques (Hidalgo et al. 2011) (Navia et al. 2013) and hand lay-up (Rodríguez et al. 2014); however, no studies were reported in the literature where tubular pieces were obtained with the technique of winding filament manufacturing.

It is for this reason that this research was manufactured a tubular piece with bio-composite material reinforced with fique fiber with the filament wilding manufacture technique, characterizing mechanically and morphologically to know its behavior this allows us to determine future potential applications.

2 Methodology

2.1 Materials

Continuous interlocking fique fibers were used as a biocomposite reinforcement, to form a 0.8 mm ± 0.02 mm diameter wire.

Polyester resin was used as a matrix for biocomposite; The resin - catalyst percentage used was 100:2 by weight.

In addition, it used a polyvinyl chloride (PVC) tube for hydraulic use, with external and internal diameter of 26,98 mm ± 0.2 mm and 21,02 ± 0.2 mm respectively, with length of 40 mm.

2.2 Composite Manufacturing

Composite materials were manufactured using the filament winding manufacturing technique; Initially the fiber was impregnated in the resin for 2 min and then bobinated on circular mold with angles of ±0°; curing was done at room temperature for 24 h.

The final fique composite geometry had an external diameter of 27 mm ± 1 mm, internal diameter of 21 mm ± 0.8 mm and a length of 400 mm, then it was sectioned in 10 specimens of 40 mm of length. Figure 1 shows the manufactured specimens.

2.3 Compression Test

The compression test was performed on MTS model C43.104 10 kN universal machine at a 2 m/min speed with a 24.2 °C temperature, using flat jaws. Five specimens were

Fig. 1. PVC and bio-composite test tubes

tested for the material, reporting the results obtained average. The tests were carried out transversely and longitudinally comparing additionally against conventional PVC tube with the same geometry. Figure 2 shows the two configurations used.

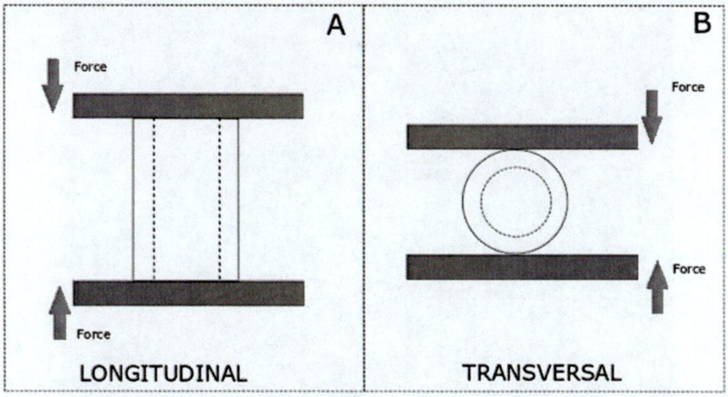

Fig. 2. Compression test configurations, A) Longitudinal; B) Transversal

2.4 Morphological Characterization

For morphological characterization, electronic scanning microscopy was used (Tescan microscope model MIRA 3 FEG-SEM with secondary electron detector model A65c SED).

Initially, values were adjusted at an acceleration voltage at 3 kV and wild field scanning mode obtaining images at few sample increases. Subsequently, the scanning mode was changed to resolution by varying the conditions to obtain images between 20 X and 500 X and at an electron acceleration voltage of up to 10 kV.

Composite materials were coated by a small layer of gold to improve electrical conductivity.

3 Results

3.1 Scanning Electron Microscopy

Figures 3 and 4 show the morphologies biocomposite at 500 magnifications in different areas. The microstructure shows gaps between the fique fibers constituents and the matrix, indicating that there was low adhesion. This is due to the hydrophobic matrix nature and the hydrophilic fibers characteristics (Muñoz et al. 2014). Results similar obtained Hidalgo et al. (Hidalgo-Salazar et al. 2013) when they made composite with fique and low Density Polyethylene-Aluminum (LDPE-Al).

Fig. 3. Electronic scanning microscopy of biocomposite material at 500X, zone 1

3.2 Compression Test

Figure 5 shows the force-displacement biocomposite versus PVC material curve in transverse compression test.

In the case of the reinforced fique fiber biocomposite, a maximum value of 0.58 kN is reached, while in the PVC pipe a 1.19 kN resistance was obtained. Regarding the materials ductility, similar behaviors were obtained, however, it should be remembered that the specimens are hollow and having the same geometry, allowed similar displacements.

Figure 6 shows the force-displacement biocomposite versus PVC material curve in longitudinal compression test.

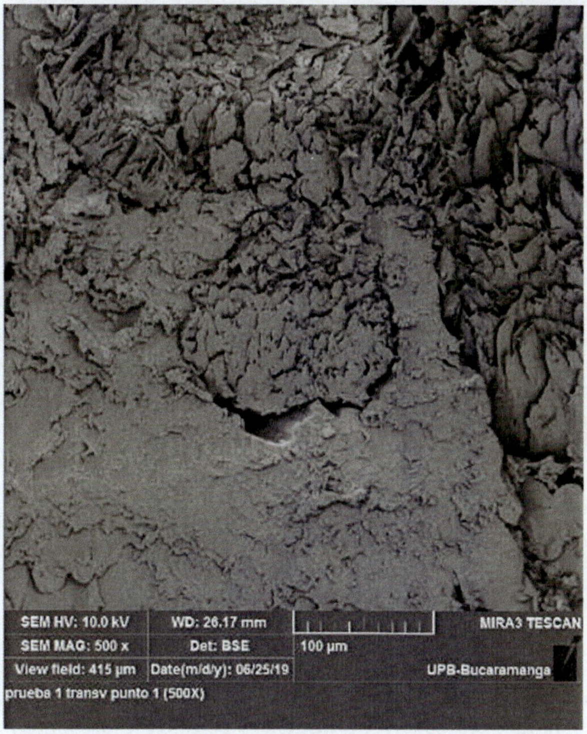

Fig. 4. Electronic scanning microscopy of biocomposite material at 500X, zone 2

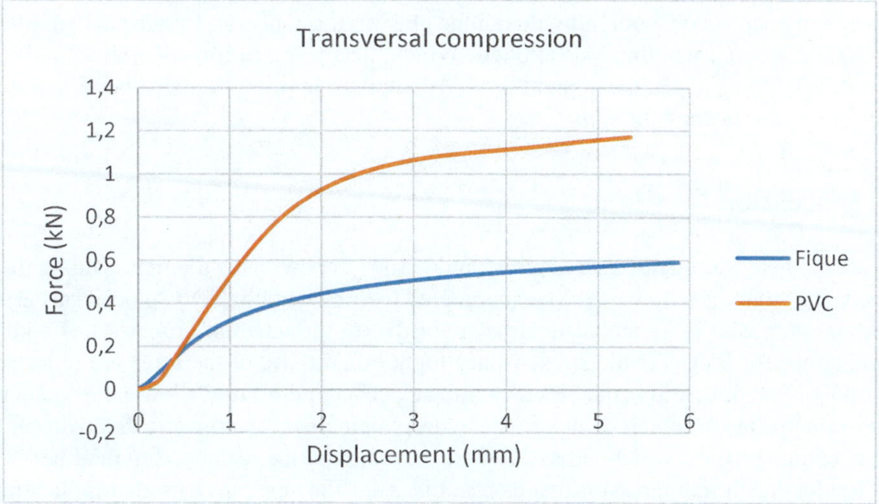

Fig. 5. Transversal compression test

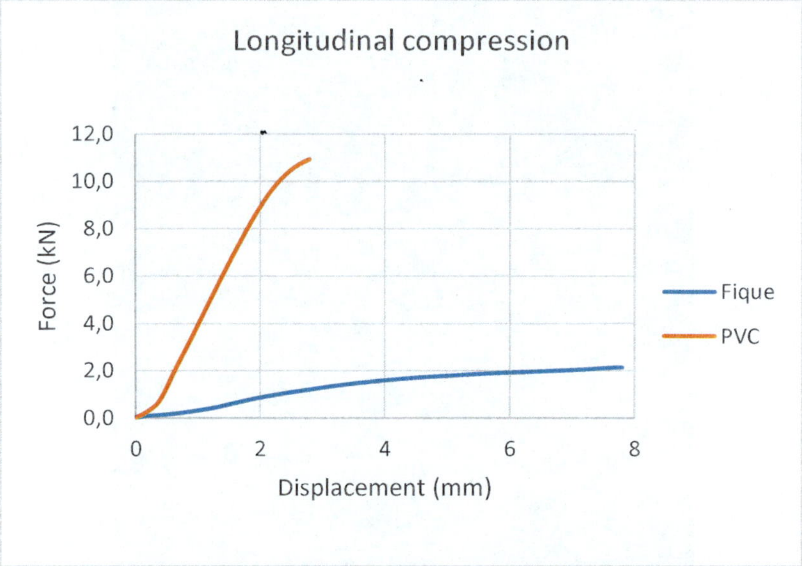

Fig. 6. Longitudinal compression test

The PVC material achieved a resistance of 11.2 kN, superior to the fique material that had a 2.1 kN resistance. However, the fique bio-composite presented a greater plastic deformation due to the spaces that were generated between fiber and fiber when winding.

As it is observed the fique composite presented a lower mechanical performance than the PVC pipe in transversal and longitudinal compression test. This was due to low matrix - fiber adhesion, poor individual fique fiber performance, and angle geometry with which the product winding was applied. Which, according to Eggers et al. (2019) the lower the winding angle, the higher the axial compressive strength since fiber orientation is closer to the loading direction.

4 Discussion

Bio-composite was manufactured with tubular fique fiber with the filament wilding manufacturing technique, however, when compared to a conventional PVC pipe, it surpassed the bio-composite in its mechanical behavior due to the inherent properties of natural fiber, (since for PVC Tensile stress values higher than those of the fique are reported), additionally as observed in the SEM scanning electron microscopy, a low adhesion in the fique composite was obtained due to the hydrophobic matrix nature and the hydrophilic fibers characteristics. which led to a low transmission of the resin load to the fiber.

Evaluating the compression results concludes that the pipe can be used in applications where large structural loads are not required, such as transport and electrical cables protection, however, it is should be supplemented by a study to assess their behavior towards temperature.

For future work it is recommended to apply a chemical treatment to the fique fiber in this way improve adhesion with the matrix also varies the angles of the winding which will be reflected in better mechanical properties.

References

Dun, M., et al.: Sisal fiber reinforced high density polyethylene pre-preg for potential application in filament winding. Compos. Part B **159**, 369–377 (2019). https://doi.org/10.1016/j.compositesb.2018.09.090

Eggers, F., et al.: Mechanical response of filament wound composite rings under tension and compression. Polym. Test. **78**, 105951 (2019). https://doi.org/10.1016/j.polymertesting.2019.105951

Gañán, P., Mondragon, I.: Fique fiber-reinforced polyester composites: effects of fiber surface treatments on mechanical behavior. J. Mater. Sci. **39**, 3121–3128 (2004). https://doi.org/10.1023/B:JMSC.0000025841.67124.c3

Hidalgo-salazar, M.A., Correa, J.P.: Results in Physics Mechanical and thermal properties of biocomposites from nonwoven industrial Fique fiber mats with Epoxy Resin and Linear Low Density Polyethylene. Results Phys. **8**, 461–467 (2018). https://doi.org/10.1016/j.rinp.2017.12.025

Hidalgo-Salazar, M.A., Mina, J.H., Herrera-Franco, P.J.: The effect of interfacial adhesion on the creep behaviour of LDPE-Al-Fique composite materials. Compos. Part B: Eng. **55**, 345–351 (2013). https://doi.org/10.1016/j.compositesb.2013.06.032

Hidalgo, M.A., Muñoz, M.F., Quintana, K.J.: Desempeño mecánico del compuesto polietileno-aluminio reforzado con agro fibras continuas de fique. Rev. LatinAm. Metal. Mat. **31**(2), 187–194 (2011)

Muñoz, M.F., Hidalgo, M.A., Mina, J.H.: Fibras de fique una alternativa para el reforzamento de plásticos. Influencia de la modificación superficial **12**(2), 60–70 (2014)

Navia, D.P., Ayala, A.A., Villada, H.S.: Determinación de isotermas de adsorción de agua en biocompuestos de harina termoplastica y fique. Biotecnología en el Sector Agropecuario y Agroindustria **11**(1), 144–154 (2013)

Rodríguez, L.J., Sarache, W.A., Orrego, C.E.: Compuestos de poliéster reforzados con fibra de plátano/banano (Musa paradisiaca) modificada químicamente. Comparación con fibra de vidrio y fique (Furcraea andina). Informacion Tecnologica **25**(5), 27–34 (2014). https://doi.org/10.4067/S0718-07642014000500005

Sapuan, S.M.: Composite materials (chapter 3). In: Sapuan, S.M. (ed.) Composite Materials, pp. 57–93. Butterworth-Heinemann, Boston (2017). https://doi.org/10.1016/B978-0-12-802507-9.00003-9

Sorrentino, L., et al.: Robotic filament winding: an innovative technology to manufacture complex shape structural parts. Compos. Struct. **220**, 699–707 (2019). https://doi.org/10.1016/j.compstruct.2019.04.055

Summerscales, J., Grove, S.: Manufacturing methods for natural fibre composites. In: Hodzic, A., Shanks, R. (eds.) Natural Fibre Composites, pp. 176–215. Woodhead Publishing, Cambridge (2014). https://doi.org/10.1533/9780857099228.2.176

Influence of Business Strategies in Achieving a Sustained Competitive Advantage

Elsa Beatriz Gutiérrez Navas(✉), María del Coral Pérez Ordoñez, and Marianela Luzardo Briceño

Facultad de Ingeniería Industrial, Escuela de Ingenierías, Universidad Pontificia Bolivariana – Seccional Bucaramanga, Floridablanca, Colombia
{elsa.gutierrez,maria.perezor,marianela.luzardo}@upb.edu.co

Abstract. Globalization and technological development have imposed a great challenge in the search for business competitiveness. This research studies the business, competitive or generic strategies that favored or disadvantaged the achievement of a sustained competitive advantage in the period 2013 to 2015 in medium and large companies of the Bucaramanga Metropolitan Area (AMB), for which we worked with a sample of 213 of the 685 companies under study, for a margin of error of 5% and a confidence level of 90%. To select the companies that achieved a sustained competitive advantage, those that exceeded the Dupont profitability of their respective national economic subsector in the three years of study were identified. The test of difference of proportions using the t-student distribution with the Minitab-v17® software, allows to conclude at 10% of significance that implementing the differentiation and approach or niche business strategies favored the achievement of a sustained competitive advantage in the companies and period under study. These results allow efforts to be focused by the local government and the business sector towards the improvement of regional competitiveness.

Keywords: Superior financial performance · Competitive strategy · Business strategy · Generic strategy · Sustained competitive advantage

1 Introduction

In 2017, Colombia ranked fifth in Latin America in the Global Competitiveness Index, with its goal being the third most competitive economy in this region by 2032 (Private Competitiveness Council 2017). Aware that the competitiveness of a country falls largely on the business fabric, and that globalization and technological development have imposed a great challenge in the search for business competitiveness, (Private Competitiveness Council and Universidad de los Andes 2017) have proposed to the country the challenge of establishing actions aimed at improving innovation and business dynamics and sophisticating and diversifying the economy. The above makes pertinent the study of competitive business strategies that favor the achievement of a sustained competitive advantage.

The competitive advantage refers to that in which a company is better compared to its rivals in the industry (Porter 1996), reflected in a profitability higher than the average of

M. T. Castañeda Galvis et al. (Eds.): ICIE 2019, LNMUINEN, pp. 66–75, 2021.
https://doi.org/10.1007/978-3-030-49370-7_8

other companies competing for the same customers (Hill and Jones 2011). A sustained competitive advantage is achieved when this performance is maintained, despite the best efforts of competitors to match or overcome it (Planellas 2016). However, it is not enough for a strategy to create value and cannot be duplicated so that it is considered to have a competitive advantage; it is necessary that it be tested in the market and show superiority over other strategies created by the competition (Álvarez 2003).

Given that achieving a competitive advantage implies satisfying the needs of consumers effectively and efficiently, offering products or services that consumers greatly appreciate for which a high price can be charged, or achieving lower costs that allow lower prices to be charged (Thompson 2012), there is a close relationship between competitive advantage and business strategies, competitive or generic. Hamel and Prahalad (1996) affirm that the future competitive advantage of a company will depend on the strategies implemented from its vision of the future, and on the way in which they identify and exploit the opportunities and manage the missing resources.

Business strategy has two major components: corporate strategy and business strategy, also called a competitive or generic strategy. While the corporate strategy asks: In which businesses should you participate? And what are the vertical and geographical limits that the company should have? The business strategy analyzes each of the company's business units and answers the questions: What does the competitive advantage of each business depend on? What is the ability to support this advantage over time? And what is the way to compete to improve the position of each business in the market? (Tarziján 2013). In this way, the competitive strategic allows each business unit to determine the type of positioning (Differentiation, cost leadership or hybrid) and the degree of segmentation (Attending to a large market or a reduced niche) to which it will bet to compete in the market.

From a bibliographic review in the Scopus database, different publications were identified in which the impact of the business, competitive or generic strategy on organizational performance and competitive advantage is evaluated, studying the implementation of strategies such as leadership in costs (Xu and Gao 2014; Sung 2015; Martínez-Simarro et al. 2015), differentiation (Punnakitikashem 2012; Tian et al. 2012; Kharabsheh and Aqrabawi 2013; Maas et al. 2014; Xu et al. 2014; Bustinza et al. 2015; Devakumar and Barani 2016; Ferraz and Gallardo-Vázquez 2016; Laari et al. 2017), the hybrid strategy, which merges the two previous approaches (Salavou 2013), in addition to the approach strategy or niche (Bakhrun et al. 2014).

The present work sought to study the business strategies adopted by medium and large companies located in the Bucaramanga Metropolitan Area that favored or disadvantaged the achievement of a sustained competitive advantage in the period from 2013 to 2015. From this date He managed to show that the companies under study that achieved a higher financial performance compared to the average of their industry in the three years evaluated were characterized by implementing differentiation strategies and approach or niche. Likewise, it was evidenced that, in 4 of the 9 economic subsectors studied, there is at least one business strategy that favored or disadvantaged the achievement of a sustained competitive advantage, which allows focusing efforts of the local government and the business sector towards of regional competitiveness.

2 Methodology

The documentary research presented has a comparative scope, and sought to answer the research question: Are there differences between medium and large companies in the Metropolitan Area of Bucaramanga (Colombia) that implemented the different business strategies that achieved an advantage competitive in the period 2013 to 2015, compared to those who did not achieve it? In this sense, the hypothesis of the existence of statistically significant difference between these two groups of companies is established.

The inclusion criteria used to determine the target population of the study were: being medium and large companies with headquarters in the four municipalities of the Metropolitan Area of Bucaramanga (Bucaramanga, Floridablanca, Piedecuesta and Girón) and having renewed the commercial registration and reported financial information to the Chamber of Commerce of Bucaramanga in 2013, 2014 and 2015. To identify the medium and large companies of the Bucaramanga Metropolitan Area, Law 905 of August 2 of 2004 of the Republic of Colombia was taken as a reference, taking into account the value of the assets to the year 2016 of the companies studied. Based on these inclusion criteria, a population of 685 companies distributed in 13 subsectors was determined (See Table 1).

Table 1. Target population and sample (Own elaboration)

Economic subsector	Population	% population by subsector	Proposed sample	Real sample	% Real sample by subsector
Real estate activities	45	6,6%	13	7	15,56%
Accommodation and food services	9	1,3%	3	6	66,67%
Banks	23	3,4%	7	7	30,43%
Wholesale trade	214	31,2%	61	42	19,63%
Construction	157	22,9%	44	20	12,74%
Consulting and architecture	31	4,5%	9	18	58,06%
Water distribution and evacuation	8	1,2%	2	7	87,50%
Livestock, hunting and forestry	61	8,9%	17	14	22,95%
Information and communication	9	1,3%	3	5	55,56%
Other dependencies	42	6,1%	12	30	71,43%
Health	35	5,1%	10	27	77,14%
Electricity supplies, gas	11	1,6%	3	10	90,91%
Transport and storage	40	5,8%	11	20	50,00%
Consolidated	685	100,0%	194	213	31,09%

It was determined to work for the study with a sample of 194 companies distributed in the 13 sub-economic sectors evaluated, using a stratified sampling with a margin of error of 5% and a confidence level of 90%. Finally, it was possible to collect information from 213 companies, managing to exceed the proposed sample for the consolidation of medium and large companies and for 9 economic sub sectors, with the exception of real estate activities, wholesale, construction, and livestock, hunting and forestry; on which the study could not be conclusive (See Table 1).

Superior performance and competitive advantage are not identical concepts, but in the strategic management literature its use is often exchanged and confused with the idea that whenever there is a superior performance there will be a competitive advantage and vice versa (Álvarez 2003). To assess competitive advantage and superior business performance, performance indicators such as return on investment, sales volume and market share have been widely used in the scientific literature, taking into account that they allow comparing firms of different sizes, even when they operate in different markets and use different accounting standards (Ynzunza and Izar 2013). In the present study The Dupont profitability index was used as the criterion under which the companies that achieved a sustained competitive advantage were identified, which can be calculated from Eqs. 1, 2 or 3:

$$\text{Dupont} = \text{Margin on Sales} \times \text{Asset Rotation} \tag{1}$$

$$\text{Dupont} = (\text{Net Profit})/\text{Sales} \times \text{Sales}/\text{Assets} \tag{2}$$

$$\text{Dupont} = (\text{Net Profit})/\text{Assets} \tag{3}$$

For the documentary research presented, the financial data reported by the companies to the Bucaramanga Chamber of Commerce and the financial indices by sector and economic subsector published in the Benchmark database were used as sources of information. From these companies were classified into two groups, according to their ability to exceed the average Dupont profitability index at the national level of the sub sector to which they belong in the three years under study: companies with and without a sustained competitive advantage in the period 2013 to 2015 (CA and WCA). Likewise, the mission and value proposal published by the companies in its institutional web portal was consulted to analyze the business strategies implemented.

To prove the existence or not of statistically significant differences between the proportion of companies with and without sustained competitive advantage that adopted each type of business strategy, the test of difference of proportions with the t-student distribution for a level of trust was calculated. 90% with the Minitab-v17® software.

3 Results

The Bucaramanga Metropolitan Area (Department of Santander, Colombia) has 685 medium and large companies that remained active and reported financial information to the Bucaramanga Chamber of Commerce in 2013, 2014 and 2015 with headquarters in the four municipalities of the area (Bucaramanga, Floridablanca, Piedecuesta and Girón), of which the sub sectors of Wholesale Trade (31.2%) and Construction (22.9%) stand out, evidencing the absence of manufacturing companies that meet these requirements (See Table 2).

A financial evaluation of the sample was performed 213 companies studied in order to identify which managed to exceed the average rate of return Dupont of economic subsector nationwide that belong in the three - year study, which allowed to classify them into two Groups: 93 companies with sustained competitive advantage (CA) in the period and 120 without sustained competitive advantage (WCA). Of the 9 subsectors in which it was able to overcome the representative sample, the performance highlights subsector Accommodation and Food Services, Consulting and Architecture, Information and Communication, Banking and Health, in which greater than or equal percentage to 50% of the companies achieved a competitive advantage sustained during the period (See Table 2).

Table 2. Analysis of companies that achieved or not a competitive advantage sustained by subsector (Own elaboration)

Economic subsector	Real sample	CA companies		WCA companies	
		#	%	#	%
Real estate activities	7	2	28,6%	5	71,4%
Accommodation and food services	6	4	66,7%	2	33,3%
Banks	7	4	57,1%	3	42,9%
Wholesale trade	42	21	50,0%	21	50,0%
Construction	20	6	30,0%	14	70,0%
Consulting and architecture	18	11	61,1%	7	38,9%
Water distribution and evacuation	7	3	42,9%	4	57,1%
Livestock, hunting and forestry	14	5	35,7%	9	64,3%
Information and communication	5	3	60,0%	2	40,0%
Other dependencies	30	12	40,0%	18	60,0%
Health	27	15	55,6%	12	44,4%
Electricity supplies, gas	10	4	40,0%	6	60,0%
Transport and storage	20	3	15,0%	17	85,0%
Consolidated	213	93	43,7%	120	56,3%

From the analysis of the value proposition and the differentiating factors highlighted in the mission statement of the companies under study, the type of positioning (Differentiation, Cost Leadership or Hybrid) and the degree of segmentation (Focus or Niche) to which each company studied bets according to the business strategy implemented. Graph 1 shows the difference between the proportion of medium and large companies in the Bucaramanga Metropolitan Area with and without a competitive advantage sustained in the period 2013 to 2015 that adopted each type of business strategy.

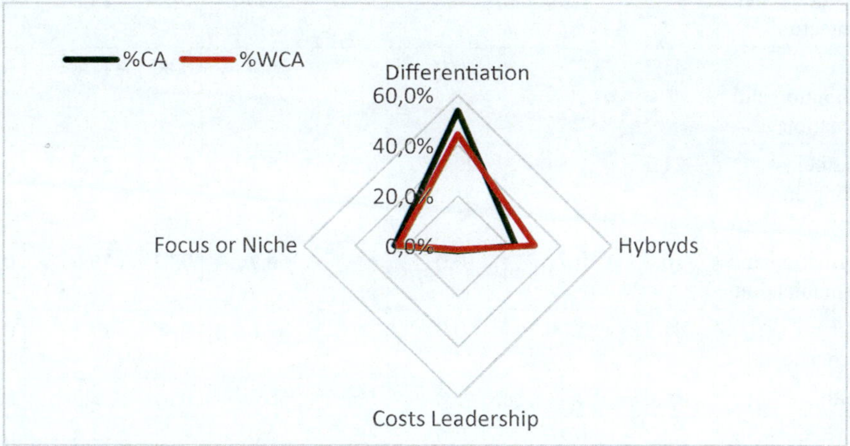

Graph 1. Comparison of the performance of companies that implement different types of competitive business or generic strategies (Own elaboration)

In Table 3 shows the difference in proportions between companies CA and WCA that implemented each type of business strategy studied (% CA - WCA), and the resulting P value of applying difference test proportions with t-distribution student, highlighting in bold the statistically significant differences at 1%, 5% and 10%, on which it is concluded.

Table 3. Comparison of the performance of companies by their sectors and type of business strategy

Economic Subsector	Differentiation		Cost leadership		Hybrid		Focus or Niche	
	% CA-WCA	P	% CA-WCA	P	% CA-WCA	P	% CA-WCA	P
Real estate activities	3,3	0,91	−10,0	0,29	23,3	0,42	−16,7	0,27
Accommodation and food services	25,0	0,25	0,0	**	−50,0	0,16	−50,0	0,16
Banks	41,7	0,23	25,0	0,25	−33,3	0,22	**83,3**	**0,00**

(*continued*)

Table 3. (*continued*)

Economic Subsector	Differentiation		Cost leadership		Hybrid		Focus or Niche	
	% CA-WCA	P	% CA-WCA	P	% CA-WCA	P	% CA-WCA	P
Wholesale trade	−3,7	0,77	−3,0	0,31	−0,6	0,96	0,0	**
Construction	12,9	0,47	7,1	0,45	−9,4	0,53	−1,8	0,87
Consulting and architecture	16,9	0,46	0,0	**	−33,8	0,10	14,5	0,21
Water distribution and evacuation	41,7	**	0,0	**	**−75,0**	**0,00**	**−17,6**	**0,06**
Livestock, hunting and forestry	−21,7	0,34	20,0	0,26	6,7	0,80	25,0	0,25
Information and communication	−16,7	0,71	0,0	**	**66,7**	**0,01**	**75,0**	**0,00**
Other dependencies	41,7	1,00	0,0	**	−2,8	0,80	13,3	0,66
Health	15,0	0,41	0,0	**	5,0	0,67	8,3	0,63
Electricity supplies, gas	33,3	0,26	0,0	**	0,0	**	−14,3	0,28
Transport and storage	**64,7**	**0,00**	0,0	**	**−41,2**	**0,00**	**19,4**	**0,15**
Consolidated	**13,6**	**0,03**	1,0	0,61	−8,1	0,14	**12,0**	**0,02**

The P-test values that could not be calculated are demarcated with **, either because no company adopted the business strategy studied generating a difference of zero proportions, or because the number of companies that adopted it was very small.

It is clarified that the value of the P test is not included when the difference between proportions is zero or the sample is very small.

The above analysis allowed the hypothesis approve that there is a positive and statistically significant difference of 13.6% between the AC and WCA companies committed to the strategy of differentiation (p-value < 0.05), according to the difference test proportions using the t-student distribution for a 95% confidence level, this strategy being decisive for the achievement of a sustained competitive advantage (See Graph 1 and Table 3). In the analysis by economic subsector (see Table 3), this hypothesis is checked only for Transport and Storage subsector (p-value 0.00).

Among the companies that are committed to the strategy of cost leadership there is a difference of 1.0% between CA and WCA, which is not statistically significant, nor for the consolidated of companies studied, nor for any of the economic subsectors (p-value $>$ 0.10), according to the test of difference of proportions using the t-student distribution for a confidence level of 90%, which shows that this strategy was not decisive for the achievement of the sustained competitive advantage.

On the other hand, there is a negative difference of 8.1% between the two groups CA and WCA firms that bet on hybrid strategy (Companies that implement the Differentiation and the cost leadership will jointly); however, this difference is not statistically significative will according to the proportions difference test using Student's t distribution for a confidence level of 90% (p-value $>$ 0.1) (See Graph 1 and Table 3). In the analysis by economic subsector, it is concluded that the hybrid strategy favors the achievement of the competitive advantage sustained in the companies of the Information and Communication subsector (p-value $<=$ 0.01), while disfavoring the subsectors of Water distribution and evacuation, and Transport and Storage (p-value = 0.00).

With respect to the degree of segmentation of the companies studied, there was a difference of 12.0% between the CA and WCA companies that implement the focus or niche strategy, directing their offer of products or services to a reduced market segment with needs very particular, statistically significant difference at 5% significance according to the test of difference of proportions using the t-student distribution (p-value $>$ 0.5). When conducting an analysis by economic subsector, it is concluded that the focus strategy favors the achievement of a sustained competitive advantage in the companies of the Banking and Information and Communication subsectors (p-value = 0.00) and disadvantages the subsector of Water Distribution and Evacuation (p-value $<$ 0.1) (See Table 3).

4 Discussion

The use of the t-student distribution allowed to approve with a level of confidence of 90% the hypothesis of the existence of statistically significant differences between the proportion of medium and large companies in the Metropolitan Area of Bucaramanga with and without sustained competitive advantage over the period 2013 to 2015 that adopted business strategies, competitive or generic differentiation and focus or niche.

Accordingly, betting on a market position based on quality and innovation attributes of products and customer service, being able to charge a higher price, becomes a source of sustained competitive advantage; while focusing on efficiency and innovation in processes to offer products or services at a low price, with basic specifications of quality and customer service, or betting on a hybrid positioning, mixing differentiation and cost leadership, was not decisive for the achievement of superior performance in the companies and period studied.

Furthermore, the study allows to conclude that serve a small market segment with particular requirements, it is critical to achieving competitive advantage held between companies and study periods. Additionally, it is concluded that, in the subsectors of Banks, Water Distribution and Evacuation, Information and Communication, and Transportation and Storage, there is at least one business strategy that favored or disadvantaged the achievement of a sustained competitive advantage.

The bibliographic review carried out by the authors allows identifying the absence of similar studies in the region, therefore, the results of this research facilitate the targeting of public and private sector actions in order to favor the conditions that stimulate business competitiveness and develop programs aimed at enhancing the sophistication of the productive apparatus of the Department of Santander (Colombia).

Extending this research to micro and small enterprises, other municipalities or departments of Colombia and other countries constitutes n into opportunities for future research, as well as complementing the study with other criteria to assess the achievement of sustained competitive advantage.

Acknowledgments. The study presented is part of the results derived from the research project "Characterization of the strategic management process in medium and large companies of the Bucaramanga Metropolitan Area that managed to develop a Competitive Advantage in the period between 2013 and 2015", (Code 026-0716-2200), sponsored by the Universidad Pontificia Bolivariana Seccional Bucaramanga (Colombia).

References

Álvarez, M.d.L: Central competencies and competitive advantage: the concept, its evolution and its applicability. Acc. Adm. **209**, 5–22 (2003)

Bakhrun, A., Anwar, S., Ramadani, L.: IT value implementation at microfinance sector. Bandung - Bali, Indonesia, sn, pp. 219–223 (2014)

Bustinza, O.F., Bigdeli, A.Z., Baines, T., Elliot, C.: Servitization and competitive advantage: the importance of organizational structure and value chain position. Res. Technol. Manage **58**(5), 53–60 (2015)

Congress of the Republic of Colombia: Law 905 of August 2 of 2004 (Colombia) By means of which Law 590 of 2000 on the promotion of the development of the micro, small and medium-sized Colombian company is modified and other provisions are issued. Congress of the Republic of Colombia, Bogotá D.C. (2004)

Private Competitiveness Council: Global Competitiveness Index 2017–2018 (2017). https://compite.com.co/indice-global-de-competitividad-2017-2018/. Accessed 25 Aug 2019

Private Competitiveness Council & Universidad de los Andes: Productivity: The key to growth for Colombia. Zetta Comunicadores, Bogotá (2017). https://compite.com.co/idc/

Devakumar, G., Barani, G.: Marketing strategies for competitive advantage: structural equation modeling approach on agricultural sector industry in South India. Int. J. Bus. Excellence **9**(2), 225–239 (2016)

Ferraz, F.A.D., Gallardo-Vázquez, D.: Measurement tool to assess the relationship between corporate social responsibility, training practices and business performance. J. Clean. Prod. **129**, 659–672 (2016)

Hamel, G., Prahalad, C.K.: Competing for the Future. Harvard Business School Press, Boston (1996)

Hill, C.W.L., Jones, G.R.: Strategic Administration, A Comprehensive Approach. Cengage Learning, Mexico (2011)

Kharabsheh, R., Aqrabawi, A.: Impact of knowledge management practices (KMPs) on competitive advantage in pharmaceutical companies, Kaunas, Lithuania, pp. 349–355 (2013)

Laari, S., Töyli, J., Ojala, L.: Supply chain perspective on competitive strategies and green supply chain management strategies. J. Clean. Prod. **141**, 1303–1315 (2017)

Martinez-Simarro, D., Devece, C., Llopis-Albert, C.: How information systems strategy moderates the relationship between business strategy and performance. J. Bus. Res. **68**(7), 1592–1594 (2015)

Maas, S., Schuster, T., Hartmann, E.: Pollution prevention and service stewardship strategies in the third-party logistics industry: effects on firm differentiation and the moderating role of environmental communication. Bus. Strategy Environ. **23**(1), 38–55 (2014)

Planellas, M.: The evolution of the strategy. Harvard Deusto (2016)

Porter, M.: Competitive advantage: creation and maintenance of superior performance. Campania Editorial Continental (1996)

Punnakitikashem, P.: Hospital quality initiative and operating efficiency: evidence from emerging economy country. Bali; Indonesia, sn, pp. 291–296 (2012)

Salavou, H.E.: Hybrid strategies in Greece: a pleasant surprise. Eur. Bus. Rev. **25**(3), 301–314 (2013)

Sung, T.K.: Application of information technology in creative economy: manufacturing vs. creative industries. Technol. Forecast. Soc. Change **96**, 111–120 (2015)

Tarziján, M.J.: Fundamentals of business strategy. Editions Catholic University of Chile, Santiago, Chile (2013)

Thompson, A.A.: Strategic Management Theory and Cases. Mc Graw Hill Interamerican of Spain, Spain (2012)

Tian, Y.Z., Li, H.T., Li, H.S., Yang, Y.: How could service-orientation contribute to business performance: the mediation effect of service practice. China, sn, pp. 513–518 (2012)

Xu, L., Gao, S., Liu, B., Grami, B., Yu, H.: Impact of the fitness between business strategy and HRM system on sustained competitive advantage. China, sn, pp. 584–589 (2014)

Ynzunza, C.B., Izar, J.M.: Effect of competitive strategies and market-oriented resources and capabilities on the growth of organizations. Acc. Adm. **58**(1), 169–197 (2013)

Preventive Measures for Ergonomic Risk Factors and Pathologies Associated with the Work of the Secretaries of the School of Engineering of Universidad Pontificia Bolivariana - Bucaramanga

Daniela Sandoval(✉), Camilo E. Rodríguez(✉), and María T. Castañeda(✉)

Faculty of Industrial Engineering, Engineering School, Universidad Pontificia Bolivariana – Bucaramanga, Floridablanca, Colombia
{daniela.sandoval.2015,camilo.rodriguez.2015,maria.castaneda}@upb.edu.co

Abstract. The present article proposes the development of a study of projective scope that had as its purpose to formulate a proposal of preventive measures, starting from the identification of the main ergonomic risk factors present in the workplaces of the secretaries of the School of Engineering of the Universidad Pontificia Bolivariana - Bucaramanga, in 2017. For this purpose, the population was censused with the Questionnaire for the Risk Factor Identification in Workplaces of the Consejo Superior de Investigaciones Científicas CSIC - Spain. The results obtained were analysed quantitatively and qualitatively. Subsequently, they were related to ergonomic measurements taken in the second half of 2016 with the REBA (Rapid Entire Body Assessment) method and instruments of this discipline. The main risk factors found were annoying reflexes and glare; repetitive movements of arms, hands and wrists; discomfort of chairs; frequent discomfort in the eyes and back; and sedentary work. Lastly, pathologies were associated to each factor, for which the proposal of preventive measures was made.

Keywords: Ergonomics · Preventive measures · Risk factors · Secretaries · Universidad Pontificia Bolivariana

1 Introduction

Ergonomics is a set of knowledge that aims to adapt products, tasks, tools, spaces and the environment in general to the capacity and needs of people. Its main objective is to improve the efficiency, safety and well-being of workers (Instituto de Biomecánica de Valencia 2014). Among the various fields of action of ergonomics is the so-called "Ergonomics of work". This type "seeks to adapt the workplace and its conditions to the person, analysing the environmental systems and their capacities" (Seguridad y Salud Premap 2017).

M. T. Castañeda Galvis et al. (Eds.): ICIE 2019, LNMUINEN, pp. 76–85, 2021.
https://doi.org/10.1007/978-3-030-49370-7_9

In the Bucaramanga section of the Bolivarian Pontifical University, one of the broadest areas that exists is administrative, whose workers are characterized by working in the same position, generally seated, for long hours. For this reason, the institution's University Welfare Office has studied and detected cases related to pain in different areas of the body, involving the administrative staff to which the twelve secretaries of the School of Engineering belong.

These discomforts, due to the areas of the body in which they occur, are related to ergonomic risk factors associated with the characteristics of the workplaces where they carry out their daily work activities. Therefore, the present research of projective scope and non-experimental design sought in the first instance to identify the main factors to which the secretaries are exposed. These are related to the development of diverse pathologies, being thus necessary the formulation of preventive measures for each one of the same ones. The fruits of this study are intended, in addition to the prevention of future injuries, to promote the main objective of ergonomics: the adaptation of work tools to human capital.

2 Methodology

The structure of the methodology is presented in four phases, which are directly related to the specific objectives proposed. In the first instance, the Questionnaire for the Identification of Risk Factors in the Workplaces of the CSIC – Spain (Consejo Superior de Investigaciones Científicas 2010) was applied. For this purpose, the census of the twelve secretaries of the School of Engineering of the Universidad Pontificia Bolivariana Bucaramanga section was used as a statistical count, in order to identify the indicators that warn of the presence of ergonomic risk factors in their workplaces. Subsequently, the most relevant ones were prioritized, relating them in turn with the ergonomic measurements carried out on the same population in the second half of 2016 through the application of the REBA method and instruments specific to the discipline studied such as: luxmeter, sonometer and thermohygrometer (Beltran et al. 2016). Thirdly, the most frequent pathologies that develop due to the existence of such risks were associated and, finally, specific preventive measures were formulated for each one of them. For this reason, the scope of the study is projective (Hurtado 2008).

3 Results

When analysing the questionnaires applied, the following results were obtained related to the most important risk factors belonging to the dimensions: Physical load and manipulation of loads, environmental conditions, special sensitivities, other ergonomic factors, work with data display screens (computers) and workplace design; together with their respective pathologies and preventive measures:

Irritating Reflexes or Glare

As shown in Fig. 1, although ten secretaries replied that there are no reflections or glare in their work environment, two of them were emphatic that there is excessive lighting in their workplaces. The secretary of Systems Engineering and graphic design wrote "I have too much light and it irritates my eyes," similarly, the graduate secretary of Industrial Engineering said "In the mornings too much light, the blind does not cover enough, passes the excessive light". It is considered a risk factor due to its descriptions; contrary to all, the secretary of Mechanical Engineering considers the light it receives to be insufficient, leading to a notorious inequality in the lighting conditions that each job has.

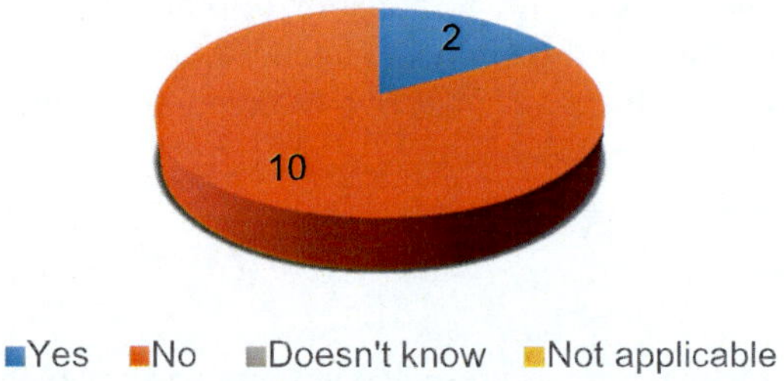

Fig. 1. Irritating reflexes or glare. Source: Authors.

The results obtained by means of the luxmeter readings (See Fig. 2) in each one of the work stations of the secretaries were taken at the beginning of the working day, therefore the values reflect the measure of the combination between natural and artificial light, where the natural light is not too much in said schedule. According to the Instituto Nacional de Higiene en el Trabajo de España (Instituto Nacional de Seguridad e Higiene en el Trabajo 2015), the minimum lighting level in areas where tasks are performed with

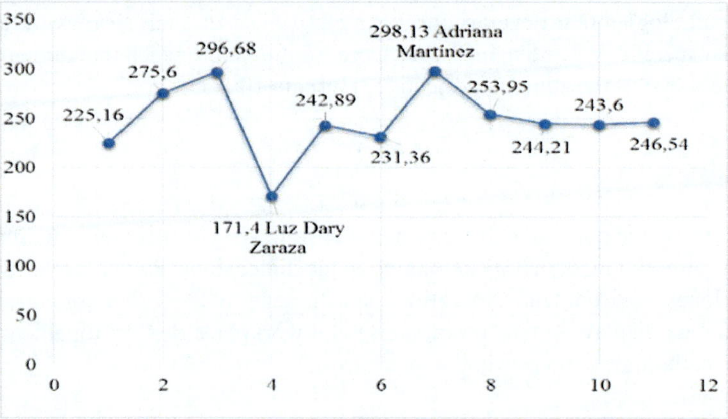

Fig. 2. Measurements obtained with the luxmeter (Lux). Source: "Ergonomic evaluation of the workplaces of the secretaries of the School of Engineering of the Universidad Pontificia Bolivariana, Bucaramanga section".

moderate visual demands is 200 lx. This allows us to deduce that the workstation of the secretary of Mechanical Engineering is insufficiently illuminated, since its value in Lux is 171.4, which coincides with its perception.

Opposite cases are those of the postgraduate secretary of Industrial Engineering, who corresponds to the measurement of 298.13, the maximum value among the entire population; and the secretary of Systems Engineering and graphic design. This last case is very specific, because the administrative was in building I in 2016, but in the present year (2017) was given a relocation to building L, which may explain the excess lighting to which it refers.

The following pathologies were associated with this risk factor, according to (Rodriguez 2014): Eye disorders, headaches and fatigue, for which they are recommended as the main preventive measures, the ones shown in Table 1:

Table 1. Disturbing glare or reflexes: preventive measures

Preventive measures
Make maximum use of natural light, which enters through windows, which must remain clean and free of obstacles. Regulate the excessive entry of light through blinds or curtains
Check that the individual lamps are properly covered so that they do not cause glare or annoying reflections on the screen
Use display screens with anti-reflective treatment or, failing that, install good quality anti-reflective filters. The filters must be cleaned regularly on both sides
Avoid leaving metal objects, glass, shiny plastics near the workplace
Do not use materials or paints whose finishes cause reflections or glare, such as glossy white walls, metals, plastics or crystals, in order to prevent annoying glare

Source: Rodríguez 2014.

Repeated Movements of Arms, Hands and Wrists

As shown in Fig. 3, nine of the twelve secretaries must perform repetitive movements of arms, hands and/or wrists. The postgraduate secretary of environmental engineering, mechanics and electronics (which is part of the affected secretaries) said "repetitive movements with the mouse". It is considered a risk factor, as it represents 75% of the population.

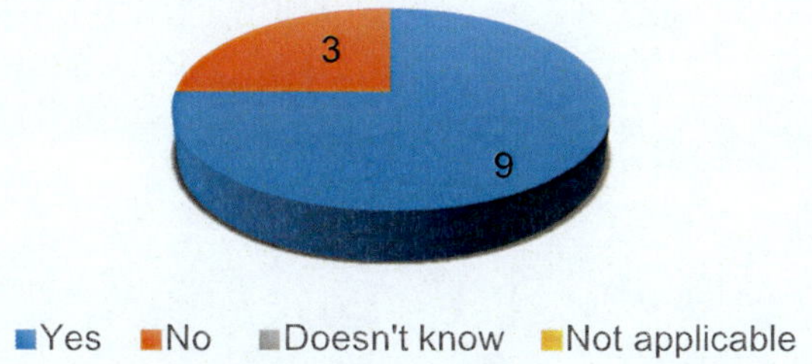

Fig. 3. Repeated movements of arms, hands and wrists. Fuente: Authors.

The results for group B (Arm, forearm and wrist) showed that ten secretaries have a low risk level, indicating a level 1, that is, action may be necessary. At the same time, the secretary of Electronic/Electrical Engineering and the secretary of Industrial Engineering (See Fig. 4) are in a medium risk represented in level 2, that is to say, the action is necessary, especially for the first of them, because according to the REBA method, there is torsion or lateral deviation in their wrists (See Fig. 5).

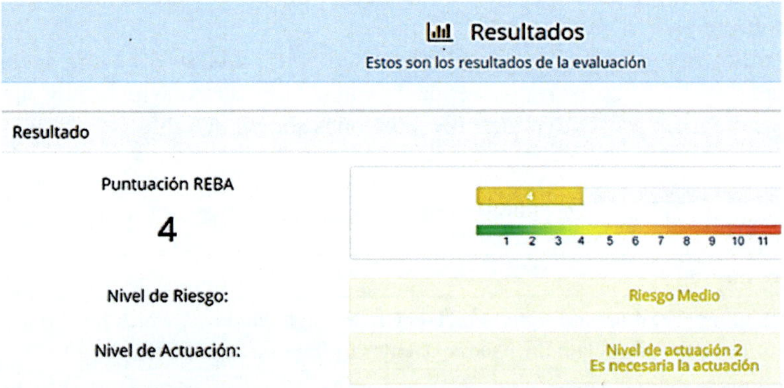

Fig. 4. Group B results of secretaries Elizabeth Barajas and Claudia Saavedra. Source: Research work "Ergonomic evaluation of the positions of the secretaries of the School of Engineering of the Universidad Pontificia Bolivariana, Bucaramanga section".

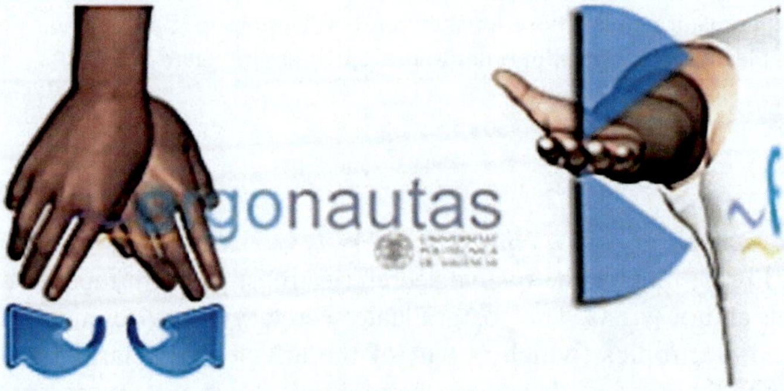

Fig. 5. Torsion or lateral deviation of the wrist of Secretary Elizabeth Barajas. Source: Research work "Ergonomic evaluation of the positions of the secretaries of the School of Engineering of the Universidad Pontificia Bolivariana, Bucaramanga section".

The following pathologies were associated with this risk factor, according to (Mutua 2012): Tendinitis/tenosivitis, Epicondylitis/epitrocleitis, Carpal tunnel/radial tunnel, Gayon Channel Syndrome, De Quervain Syndrome and Spring Finger, for which they are recommended as the main preventive measures, the ones shown in Table 2:

Table 2. Repeated movements of arms, hands and wrists: preventive measures

Preventive measures
Place your arms on the armrest so that your forearms, wrists and hands are aligned in a straight line
Avoid repetitive movements with only one arm, try to move both in an opposite or symmetrical way
In order to promote blood circulation and the mobility of the hands and fingers, the following exercise can be carried out: Extend your arms and slowly clench your fists, then open them and separate your fingers
The keyboard must act as a hand rest
When using the mouse, place it on the left or right side of the keyboard, it is recommended to have the whole arm, shoulder and wrist to move the mouse

Source Authors. Based on: Studies of musculoskeletal injuries in the working environment of pimes (Fundación para la Prevención de Riesgos Laborales 2014)

Chair Discomfort

As shown in Fig. 6, of the twelve secretaries, eight replied that the chair does not represent discomfort. However, the remaining four secretaries presented dissatisfaction with the work chair, where the secretary of Systems Engineering and Graphic Design said: "The back of the chair is uncomfortable" and the secretary of Environmental Engineering "It would be good to provide a relaxing lumbar on the chair part of the back, since our work in large percentage we are sitting". Because of his observations, this was taken into account as an indicator of risk factor.

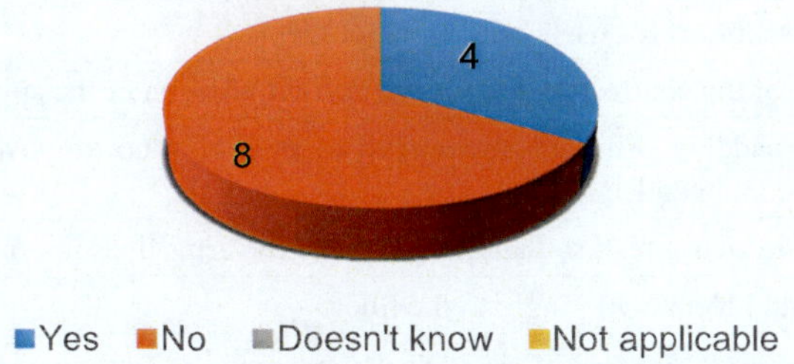

Fig. 6. Chair discomfort. Source: Authors.

The results for group A (trunk, neck and legs) showed that 9 secretaries have a low level of risk, indicating a level 1, i.e. action may be necessary. However, the remaining 3: The undergraduate secretary of Electronic/Electrical Engineering (detected with medium risk level), the postgraduate secretary of Electronic and Environmental Engineering and the undergraduate secretary of Industrial Engineering present a medium risk level, for which an action level 2 is indicated, being necessary (See Fig. 7).

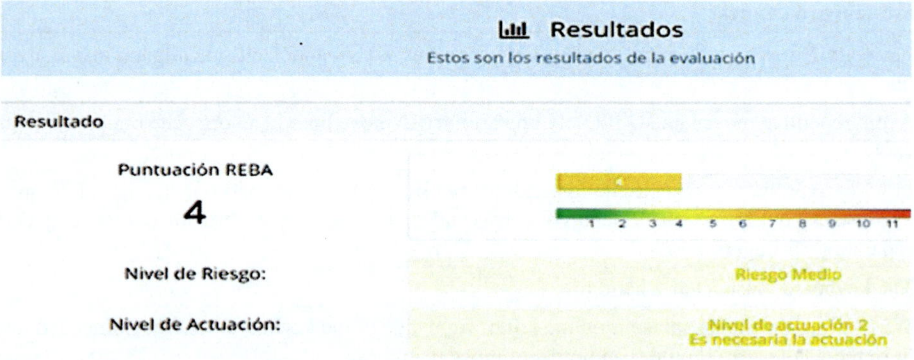

Fig. 7. Group A results of secretaries Elizabeth Barajas and Claudia Saavedra. Source: Research work "Ergonomic evaluation of the jobs of the secretaries of the School of Engineering of the Universidad Pontificia Bolivariana, Bucaramanga section".

The following pathologies were associated with this risk factor, according (El Universal 2017) and (Cuidado de la Salud 2015): Dorsalgia, Cervicalgia, Kyphosis, Torticollis, Low back pain and Myopathies, for which they are recommended as the main preventive measures, the ones shown in Table 3:

Table 3. Chair discomfort: preventive measures

Preventive measures
Adjust the depth of the chair according to the needs of each secretary. A very deep chair, generates greater discomfort for users with small context
The lumbar support of the chair should favor the upright position of the spinal column
A chair with excess padding will allow the user to adapt to poor posture over time. A 2 cm deep upholstery is recommended
The chair should have an armrest, so that the weight of the arms does not fall on the upper back
Ideally, the legs should form a 90° angle to the floor

Source Authors. Based on: Studies of musculoskeletal injuries in the working environment of pimes (Fundación para la Prevención de Riesgos Laborales 2014)

Frequent Discomfort in the Eyes, Back, Wrists, Among Others
As shown in Fig. 8, it took into account the discomfort in the eyes, back, wrists, among others. Seven out of twelve secretaries perceive discomfort frequently, being thus approximately 58% of the affected population. Two of them made their personal observations, the secretary of Electronic Engineering: "Discomfort in the back", the secretary of Systems Engineering and Graphic Design: "Discomfort in the eyes and back". Regarding the frequent discomfort in the eyes, no measurements are presented. As for the frequent back discomfort, its analysis is presented in the previous indicator "Chair discomfort" and in the wrists in the indicator "Repetitive movements of arms, hands and wrists".

Fig. 8. Frequent discomfort in the eyes, back, wrists, among others. Source: Authors.

The following pathologies were associated with this risk factor, according (Josantonius 2010): Glaucoma, Scoliosis, Tendinitis, Conjunctivitis, Dorsalgia, Carpal tunnel syndrome, Diplopia or double vision, Low back pain and Arthritis, for which they are recommended as the main preventive measures, the ones shown in Table 4:

Table 4. Frequent discomfort in the eyes, back, wrists, among others: preventive measures

Preventive measures
Try to have a visual distance of 60 cm. Thus, a clear visualization of the text can be obtained on the computer screen
Working with objects that do not generate reflection on the surface of the face or do not contain a shiny surface, thus avoiding discomfort to the eyes
At the end of the workday, each secretary should devote 5 min to rest the eyes. A couple of examples to apply them are: Press the eyebrows, with the help of the index fingers and thumb in a circular way with repetitions of 3 to 5 times and with the eyes closed, make upward and downward movements
The twelve administrative chairs should have a comfortable, soft backrest intended to support the lower back. In addition, the contours of the vertical part of the seat should be aligned with the curve of the lower part of the spine
Position the computer keyboard at an angle of no more than 25° so that the twelve secretaries obtain a wrist position in a straight line
Avoid wrist flexions while the secretaries are in constant use of the keyboard

Source Authors. Based on: Studies of musculoskeletal injuries in the working environment of pimes (Fundación para la Prevención de Riesgos Laborales 2014)

Sedentary Work
As shown in Fig. 9, all the secretaries stated that they work in a sedentary manner, in accordance with the requirement that their work demands of them. Therefore, it is considered an important risk factor, for which the related measurements obtained are the same as those presented in Sects. 2 and 3.

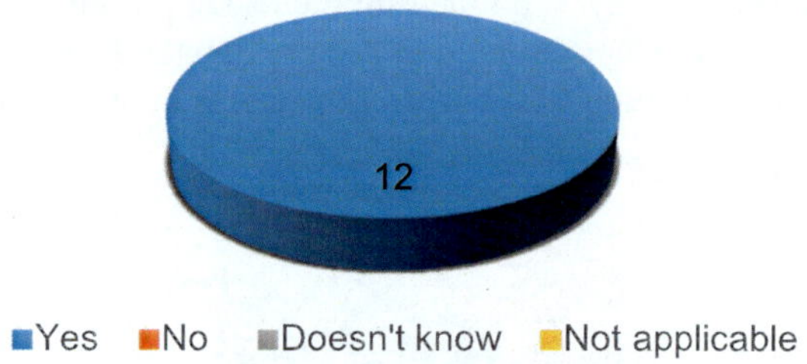

Fig. 9. Sedentary work. Source: Authors.

The following pathologies were associated with this risk factor, according (Claire 2014): Obesity, Diabetes, Fibromyalgia, Heart problems, Muscle problems and Bone problems, for which they are recommended as the main preventive measures, the ones shown in Table 5:

Table 5. Sedentary work

Preventive measures
Change posture often during the day
Take active breaks on a regular basis
Getting up to talk to co-workers instead of sending them an e-mail or text message
If mobilization is necessary, use the stairs, not the elevator
Adjust the height of the chair so that the shoulders are relaxed and the elbows have a comfortable support point
Distribute weight evenly throughout the chair, using the entire seat and back to support the body
Feet should rest firmly on the floor when sitting
Do not sit with your legs crossed, as it causes deviations in the spine and hinders blood circulation in the legs

Source Authors. Based on: Studies of musculoskeletal injuries in the working environment of pimes (Fundación para la Prevención de Riesgos Laborales 2014)

4 Discussion

The results obtained from this research work generate concern about some of the physical conditions of the jobs of the secretariats of the School of Engineering of the Bolivarian Pontifical University Bucaramanga section. It is considered advisable that the staff of the

University Welfare Department carry out an analysis of them together with experts in the subject of Occupational Risk Management (ARL), with the objective of implementing improvement plans in addition to offering more adequate conditions and, to a certain degree, personalized in the position of each affected secretary. Similarly, socialize and propose frequent active breaks, considering that the most relevant risk factor is sedentary work during the working day; and other preventive measures specifically proposed for each factor, in order to avoid in the medium term ergonomic risks and the possible development of pathologies associated to which administrative risks are exposed due to the characteristics of the site where they carry out their daily occupational activities.

References

Instituto de Biomecánica de Valencia, Introducción a la ergonomía (2014). http://ergodep.ibv.org/documentos-de-formacion/1-documentos-de-introduccion/501-introduccion-a-la-ergonomia.html

Hurtado, J.: El proyecto de investigación (2008). https://es.scribd.com/document/325498589/EL-PROYECTO-DE-INVESTIGACION-HURTADO-2012-pdf

Beltran, J., et al.: Valoración ergonómica en los puestos de trabajo de las secretarias de la Escuela de Ingeniería de la Universidad Pontificia Bolivariana seccional Bucaramanga, Bucaramanga: s.n (2016).

Consejo Superior de Investigaciones Científicas: Cuestionario para la identificación de factores de riesgo en los puestos de trabajo del CSIC (2010) https://www.icv.csic.es/prevencion/Documentos/manuales/cuestionario.pdf

Seguridad y Salud Premap: Ergonomía en el puesto de trabajo (2017). http://www.prevencionfremap.es/prevencion-ergonomia.php

Instituto Nacional de Seguridad e Higiene en el Trabajo, Iluminación en el puesto de trabajo (2015). https://www.insst.es/InshtWeb/Contenidos/Documentacion/Iluminacion%20en%20el%20puesto%20de%20trabajo.pdf

Rodriguez, P.: Efectos de la iluminación inadecuada en la salud (2014). https://elnacional.com.do/efectos-de-la-iluminacion-inadecuada-en-la-salud/

Mutua, M.: Definición por lesión de movimiento repetitivo (2012). https://www.maz.es/MutuaMAZ/Paginas/ConoceNuestraMutua.aspx

El Universal: Cinco enfermedades ocasionadas por una mala postura en el trabajo (2017). https://www.eluniversal.com.mx/articulo/ciencia-y-salud/salud/2017/03/1/cinco-enfermedades-ocasionadas-por-una-mala-postura-en-el

Josantonius: El sentido de la vista; Enfermedades, defectos y problemas (2010). http://josantonius.blogspot.com/2010/12/el-sentido-de-la-vista-enfermedades.html

Claire, M.: Cinco enfermedades relacionadas con el sedentarismo (2014). https://www.marie-claire.es/belleza/fitness/fotos/5-enfermedades-relacionadas-con-el-sedentarismo/problemascardiacos

Cuidado de la Salud: Enfermedades causadas por una mala postura al sentarse (2015). http://www.cuidadodelasalud.com/f-as/ii-en/2-ir/enfermedades-causadas-por-una-mala-postura-al-sentarse/

Fundación para la Prevención de Riesgos Laborales, Estudio de las lesiones muscoloesqueléticas en el ámbito laboral de las pimes (2014). https://www.asemprevencion.es/blog/manuales-guias/estudio-de-las-lesiones-musculoesqueleticas-en-el-ambito-laboral-de-las-pimes

Ergonomic Methods Evaluation at a Service Station: Bibliometric Analysis and Review

Marlon S. Ramírez[1](✉), Silvia D. Arciniegas[1](✉), Maria M. Muñoz[1], Alejandro Acevedo[2](✉), Duwamg A. Prada[3](✉), and Jairo Nuñez[1](✉)

[1] Facultad de Ingeniería Industrial, Escuela de Ingenierías, Universidad Pontificia Bolivariana – Seccional Bucaramanga, Floridablanca, Colombia
{marlon.ramirez.2016,silvia.arciniegas.2016,maria.munoz.2016, jairo.nunez}@upb.edu.co

[2] Facultad de Administración de Empresas, Universidad Pontificia Bolivariana – Seccional Bucaramanga, Floridablanca, Colombia
alejandro.acevedoa@upb.edu.co

[3] Jefatura Departamento de Ciencias Básicas, Universidad Pontificia Bolivariana – Seccional Bucaramanga, Floridablanca, Colombia
duwamg.prada@upb.edu.co

Abstract. The purpose of this study is to review the literature on possible ergonomic methods to evaluate in a service station, this methods are known as the Rapid Entire Body Assessment (REBA), the Ovako Working Posture Analysis System (OWAS) and the Nordic Questionnaire, which were determined given that they recognize musculoskeletal disorders and corporal affectations in all the limbs of the worker. In this sense, a bibliometric and systematic analysis of the literature was carried out where the study variables, most representative authors, frequency of key words and geographic concentration were determined.

Keywords: Bibliometric analysis · Ergonomics · Nordic questionnaire · OWAS · REBA

1 Introduction

At the time, ergonomics it is considered as a multidisciplinary field study of interest, which looks for health problems prevention who affects the worker lifestyle, in that way, the security, efficiency and comfort on the workplace it is been maximizing. So that the ergonomics requires a series of disciplines such as experimental psychology in the study of attitudes and the rest of human factors, the medicine and work physiology so the human body reactions can be analyzed, the biometrics and biomechanics which studies the postures and the movement during the work and does the job analysis to know the process, charges and it distribution inside the system. (Ramirez Cavassa, 2000) (Oborne, 1987). In this way, an ergonomic study is carried out through various "ergonomic evaluation methods" because these ones allowed to recognize musculoskeletal disorders and bodily affectations in the worker extremities.

M. T. Castañeda Galvis et al. (Eds.): ICIE 2019, LNMUINEN, pp. 86–101, 2021.
https://doi.org/10.1007/978-3-030-49370-7_10

Because the ergonomic study it is going to be specifically applied to employees of liquid hydrocarbons service station, it is necessary that the actual investigation and bibliographic review, it is focused on the definition, interpretation, and application of the ergonomic methods known as Rapid Entire Body Assessment (REBA), Ovako Working Analysis System (OWAS) and the Nordic Questionnaire; all of those are observational procedures about the different postures adopted by the worker, with the final purpose to opt for practical solutions to problems or specific risk agents turning in corrective actions that can help to improve the workplace.

2 Objectives

- To review the literature about possible ergonomic methods (OWAS, REBA and Nordic Questionnaire) to evaluate the employees of a liquid hydrocarbons service station.
- To Synthesize the information on application of the ergonomics methods through the scientific articles selection.
- To expose a bibliometric analysis of the literature according to studies variable, most representative authors, key words frequency and geographic concentration.

3 Methodology

The current investigation is exploratory type because it is based on the literature, whose function is to use information of referenced investigation of scientific articles, which are supplied for a specific data base, Scopus and Web of Science, to know of a more detailed way the links around the ergonomic, through some ergonomic evaluation methods on the work place, such a Rapid Entire Body Assessment (REBA), Ovako Working Posture Analysis System (OWAS) and Nordic Questionnaire, which as a result it will present the ergonomic behavior, the application field, the publication frequency concerning to the topic, to give the support and orientation to the application at an oil service station.

The first investigation phase was the systematic literature review which include three stages: planning, implementation and report. On the planning, it was established the review needs, analyzing the proposal of possible ergonomic methods applied on the Service Station (REBA, OWAS and Nordic Questionnaire) and finally, the investigation protocol was developed. On the implementation stage, the investigation material was determined, the studies where selected in accordance to the quality criteria with the purpose of synthesize the information on application of those ergonomic methods. Finally, the report it is about of the presentation of information, through a bibliometric analysis and principal results, where the study variables were determined, the most representative authors, key words frequency and geographic concentration.

4 Results

Phase 1. Literature systematic review

For this phase, the investigation planning was made, where the proposal of the possible ergonomics methods to applied on a service station was analyzed, these must to be related with standing work or with the evaluation of musculoskeletal disorders on all the body parts.

To obtain the investigation material, two data bases were explored, Scopus y Web of Science, for this the articles number resulting and each thematic of interest was taken in to account. About the search equation, it was precise according the intersection and relation between the convenience topics, stipulate the documents were define on the title. With the search equation presented on Table 1 a total number of 345 documents were obtained for REBA method, 311 for OWAS and 564 for the Nordic Questionnaire. Of which, through an elimination and depuration process, cleaning criteria which corresponds to the title analysis that have no relation at all with ergonomics were implemented, also it was excluded through a detailed observation on the abstract of each article, those who only apply the methods in the superior parts of human body. Finally, 98 results to analyze for the REBA evaluation method, 111 for OWAS and 124 for the Nordic Questionnaire. This search product registry the results before June 2019.

Table 1. Search equation

SCOPUS	WEB OF SCIENCE
REBA	
Search Eq (Rapid Entire Body Assessment or reba) = 9/06/2019	
264	90
Total titles applying cleaning criteria such as duplicates and no registry 345	
OWAS	
Search Eq. (Ovako Working Analysis System OR OWAS) 9/06/2019	
296	36
Total titles applying cleaning criteria such as duplicates and no registry 311	
CUESTIONARIO NÓRDICO	
Search Eq. (("nordic question*")) 21/06/2019	
488	349
Total titles applying cleaning criteria such as duplicates and no registry 564	

Bibliometric Analysis

For the bibliometric analysis, some variables were analyzed: selected articles affiliations, those who has the biggest relevance, which means, the articles that have been most cited as a support to investigations related with ergonomic, health care, security and comfort; the authors that who has written with most frequency in accordance with the evaluation method in the workplace. As a result of the variables previously mentioned, it allowed to relate knowledge networks between authors, key words, country relation, years of publication and all the support that help to transform information into knowledge, using the text mining tool Vantage Point, which explores, recollects and organize the information in a proper way in accordance with the results of the present.

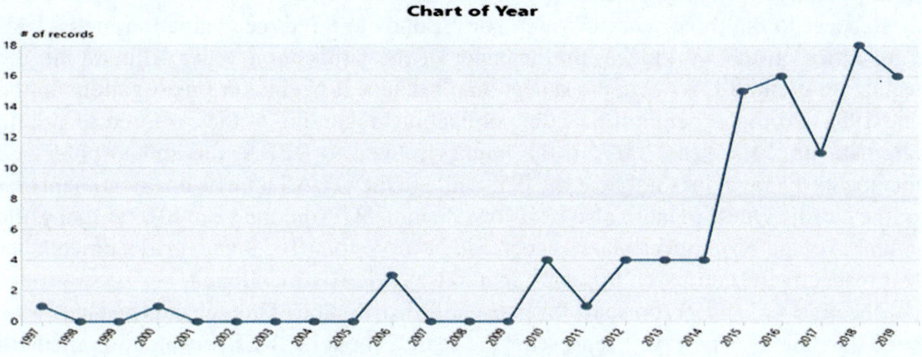

Fig. 1. Annual behavior REBA

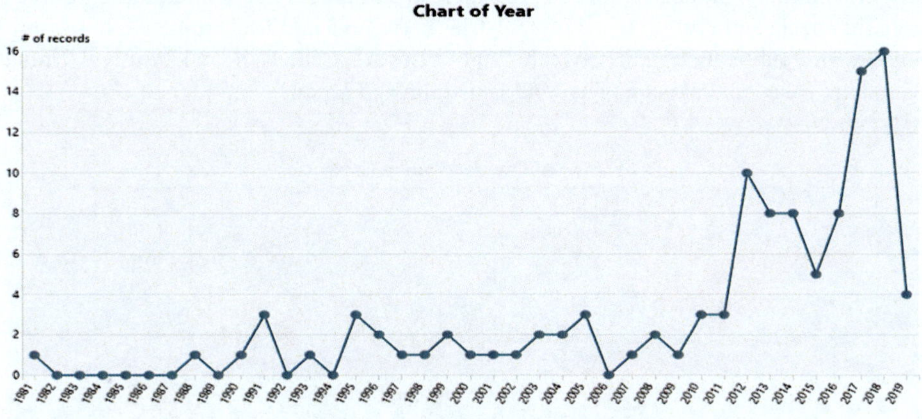

Fig. 2. Annual behavior OWAS

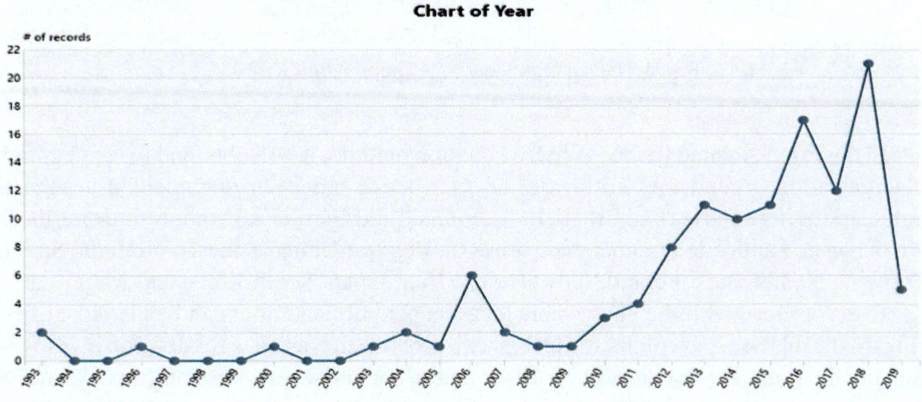

Fig. 3. Annual behavior nordic questionnaire

Respect to the three search, which corresponds to the three evaluation methods of ergonomics on the workplace, the behavior of the publication years differs a bit, the evaluation method OWAS is the oldest one, because it begins on investigations in the year 1981. From the beginning of the publications regarding to the explored search on data base until the year 2009, in the inquiry related to REBA, the annual published documents average does not beat the 0,5; while in the OWAS it beats one document and for the Nordic Questionnaire also beats one document. From the year 2010 to today, the annual average of publications has increase in more than 10, 9 and 11 documents per year respectively, compared to search methods previously mentioned.

The Figs. 4, 5 and 7 represents the geospatial distribution of the selected and later analyzed documents. The Fig. 4 represents the search about the REBA evaluation method, where the countries of India and Malaysia are the ones who has documented and publish more than 11 documents related with ergonomics and the evaluation method previously mentioned, on the second category there are countries like United States, Turkey, Iran, Indonesia and Philippines with registries between 5 and 10 documents. In the third place, with a publication rank between 2 and 4 titles are, Canada, Brazil, Portugal, United Kingdom, Sweden, Pakistan, Thailand, Australia and Japan.

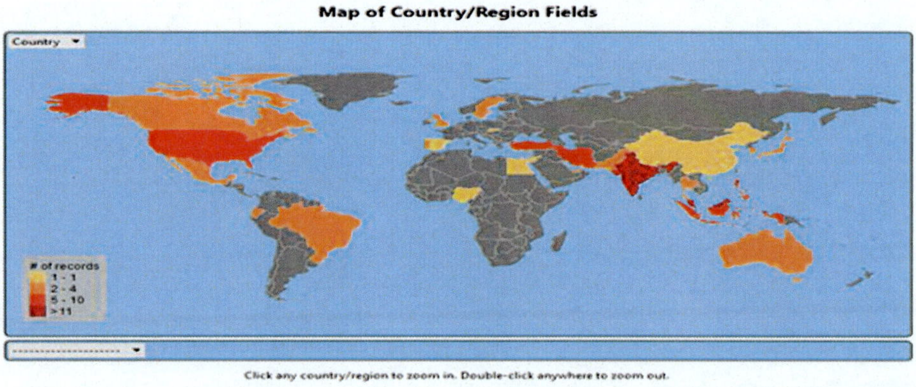

Fig. 4. Geospatial location of publication REBA

In the Fig. 5, related to the OWAS evaluation method, it was obtained two categories of publication around the world, on the first one it has been documented around 8 publications in countries like Brazil, India, Finland and Germany. In the second one, there are between 3 and 7 documents, here are countries like United States, Colombia, Spain, Portugal, Netherlands, Poland, Italy, Turkey, Iran, Japan, South Korea and Malaysia.

In accordance with the geographic location per publication, it can be also identified (Fig. 6) the relation and collaboration that exists between countries for develop investigation publications, it was determined 4 association networks between countries: Australia and Italy, South Korea and United States, Japan and Iran, and Spain with Portugal and Brazil. Also, it can be determined that countries like India, Germany and Finland has a big activity of publications to take studies of repetitive evaluation as the OWAS.

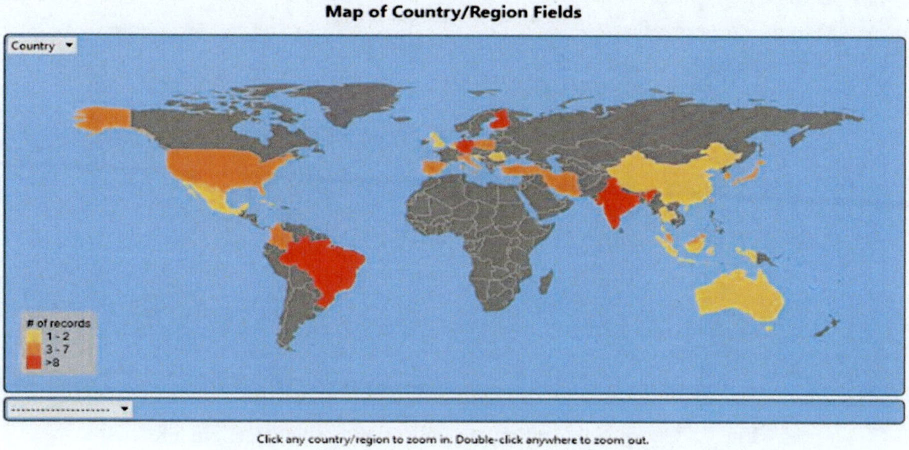

Fig. 5. Geospatial location of publication OWAS

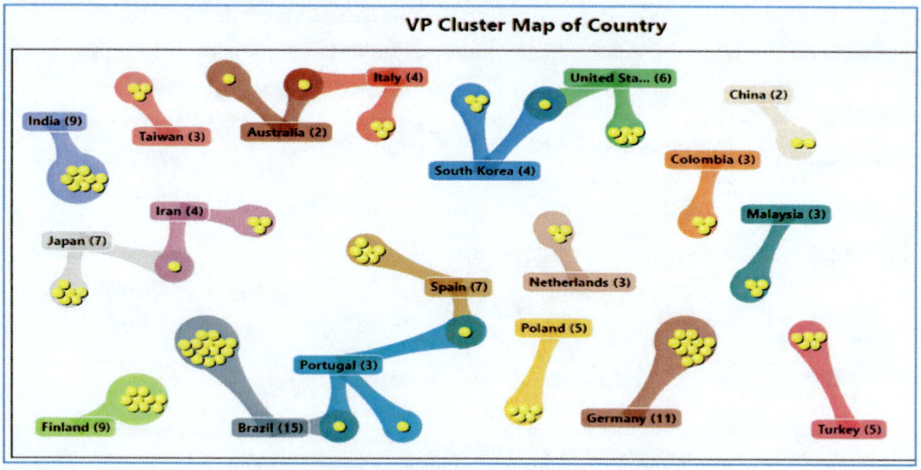

Fig. 6. Relation between countries OWAS

In the Fig. 7 it was mainly registered three categories, the first one agglomerate countries like India and Iran which have more than 12 publications about the study and application of the Nordic Questionnaire. On the second category, there are those countries that have registries between 5 and 11 documents such as Brazil, United States, Sweden, Malaysia and Philippines. On the third category, there are countries like Canada, Colombia, Spain, France, United Kingdom, Nigeria, Greece and Estonia, registering between 2 and 4 publications.

Also in the Fig. 8 it was found three collaboration networks between countries, one of them is Ireland and France, the other one Germany with Sweden and the last one a relation between Iran, Spain, United States and Brazil. It is notably that countries like India, Malaysia and Iran have a high number of publications, which supposed a great mastery and management of the Nordic Questionnaire.

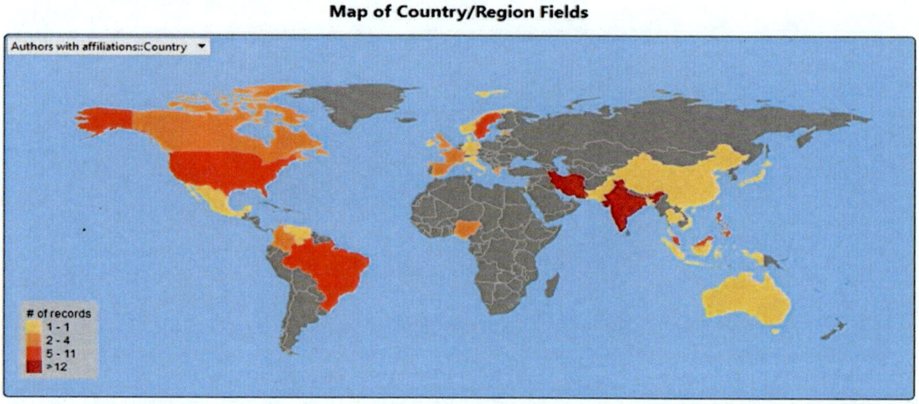

Fig. 7. Geospatial location of publication nordic questionnaire

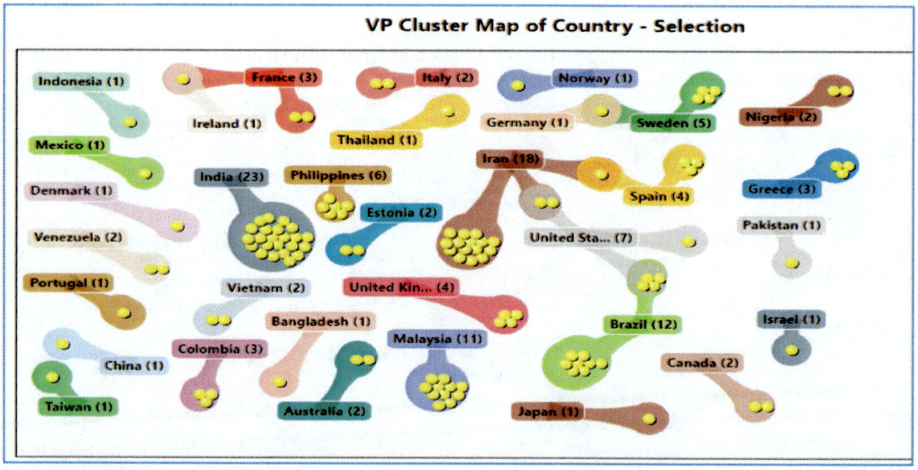

Fig. 8. Relation between countries nordic questionnaire

The Figs. 9, 10 and 11 allowed to identify the knowledge networks between authors, where the REBA evaluation method have 5 networks, OWAS 8 and the Nordic Questionnaire 8; The symbol size that represents each author means the publication frequency that exist until June 2019 and the union between authors symbolizes the relation and collaboration between them, this allowed to synthetize the information to do more specific studies.

The Fig. 9 mainly analyses the greater number of members formed by Mushtaq F, Shah Z.A, Sheikh I.A, Ashraf M and Amjad A, affiliates to the Institute of quality management and technology, in Punjab University, Pakistan. They studied the risk factors that takes to musculoskeletal problems because uncomfortable postures on the manufacture of clothes in the Pakistan industry. Also, authors with the most publication frequency in this evaluation method are taking into account, one of them is Deros B.M., researcher

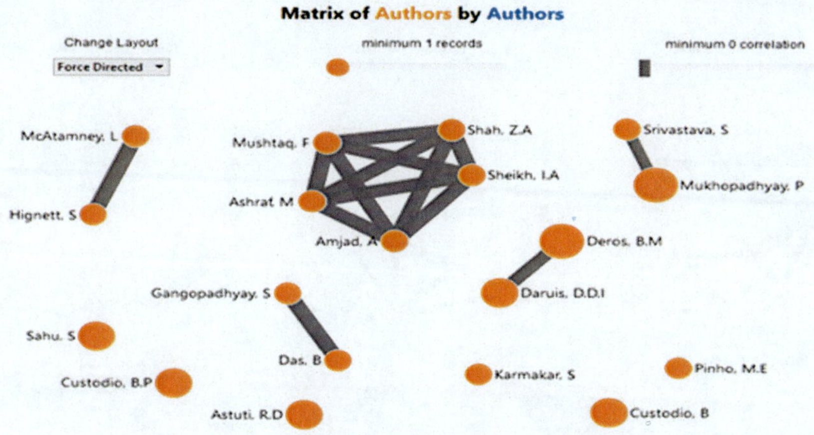

Fig. 9. Knowledge network REBA

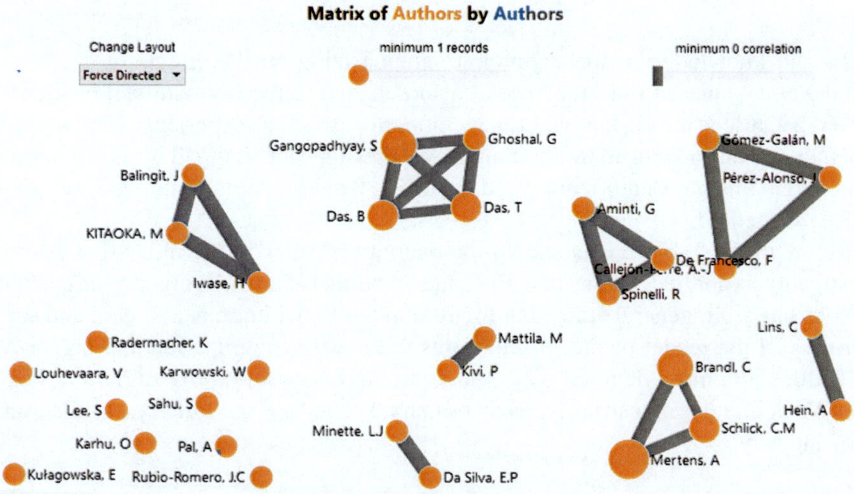

Fig. 10. Knowledge networks OWAS

of the mechanics and materials engineering of the National University of Kebangsaan, Malaysia, who based his scientific documents on the human being postures at the industry which causes musculoskeletal disorders. And Mukhopadhyay P, a professor of industrial ergonomic in the Indian Institute of design and manufacture of information technology at Jabalpur. This researcher is focused on ergonomic design problems inside various industrials sectors with the purpose of reduce the musculoskeletal injuries for the health and security.

In accordance to the Figs. 10 and 11, the biggest knowledge network it is analyzed, notably, the Fig. 10 which represents the OWAS evaluation method has the same authors as the Fig. 11 which represents the Nordic Questionnaire evaluation method. In this networks, the most representative researcher is Gangopadhyay S., who integrate all his

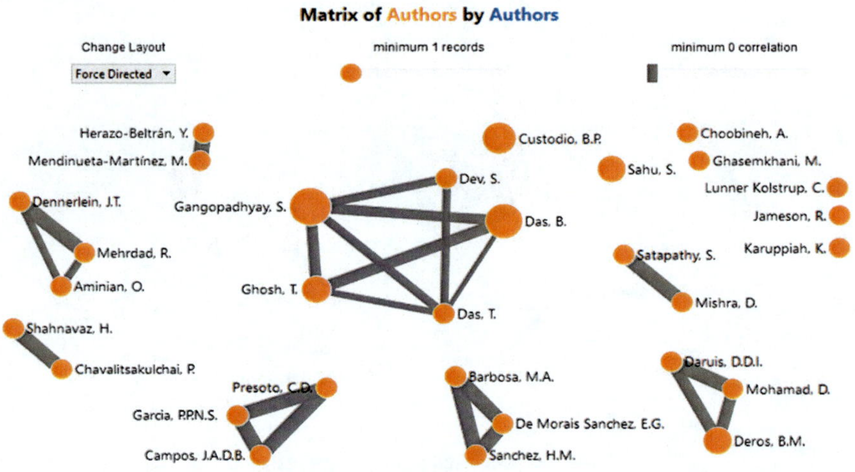

Fig. 11. Knowledge networks nordic questionnaire

efforts and knowledge in the ergonomic, emphasizing on the effects of the postures with the emergence of musculoskeletal disorders and in the occupational health of the worker, because is a well-known Indian ergonomist who has experience both in the first level industry and also in an international level prestige University. It is also important to mention that this same author appeared in the Fig. 9 in a less representative way applying the REBA method.

Below in the Table 2 it is shown the quantity of titles in the industries where the investigation is applied, thanks to it, the understanding is that these ergonomic evaluation methods has a big general study, but focused in the agriculture, health care and service industries. If the reader of this article wants to know the aggrupation and organization of the titles in a more detailed way, please feel free to write to the authors through e-mail. We define the industrial types as primary, secondary and tertiary, detailed on the following way:

- On the primary sector, were considerated studies associated to natural resources extractions, agriculture, as an example.
- The secondary sector, were considerated studies on the transformation of processed products, for, Example the Automotive industry, among others.
- The third sector to the service supply to thirds enterprises, as an example the medical services among others.

Table 2. Titles location in industry

Industrial sector	OWAS	REBA	NORDIC
Primary sector	25	20	27
Secondary sector	18	27	22
Third sector	28	22	52

Comparing the three searching methods represented in the Table 3, with the respective selection of relevant investigations, analyzing the lower parts or the entire parts of the human body, in addition to the identified methods, some documents support it study next to the Rapid Upper Limb Assessment method (RULA), which measurement consists in evaluate the musculoskeletal disorders of the upper body parts, based on the assessment of the postural load. 25 titles of REBA evaluation method, 16 of OWAS and 15 of the Nordic Questionnaire evaluates jointly the RULA method.

Table 3. Comparison between search topics.

REBA	OWAS	NÓRDICO	body	RULA
98	111	124		25 16 15

The Figs. 12, 13 and 14 presents the affiliations where investigations about the evaluation methods REBA, OWAS and Nordic Questionnaire were made. The number that appears in each bar chart represents the frequency of study publication by institution. The affiliation inside a document has the purpose to support the study developed by the researcher, besides having used the infrastructure for it elaboration. In the affiliations which corresponds to REBA, the two most featured institutions were the Philippines University and the University of Calcutta, the first is one of the 100 best Universities of Asia and the second is the fifth best University of India. The Calcutta University has an ergonomic laboratory. For the OWAS, the technique University of Aquisgran a German institution known internationally because has a great developed in the engineers, computing, physics and chemistry, and again the Calcutta University directly from the science, technology and agriculture faculty. Finally, concerning to the Nordic Questionnaire, the Tehran University is an Institution with the first place in the ranking of investigations institutes from Iran until the year 2017 and again, the Calcutta University with 8 publications concerning to the Nordic Questionnaire. It is important to mention that the institutions with the most input to the ergonomic study can be find in the Asian continent.

Fig. 12. REBA affiliations

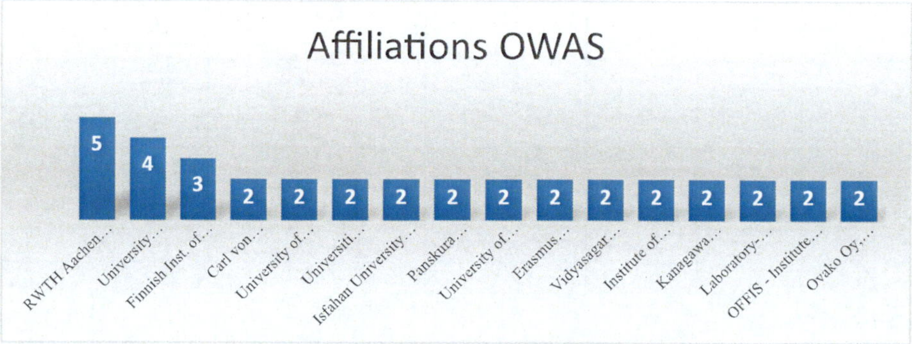

Fig. 13. OWAS affiliations

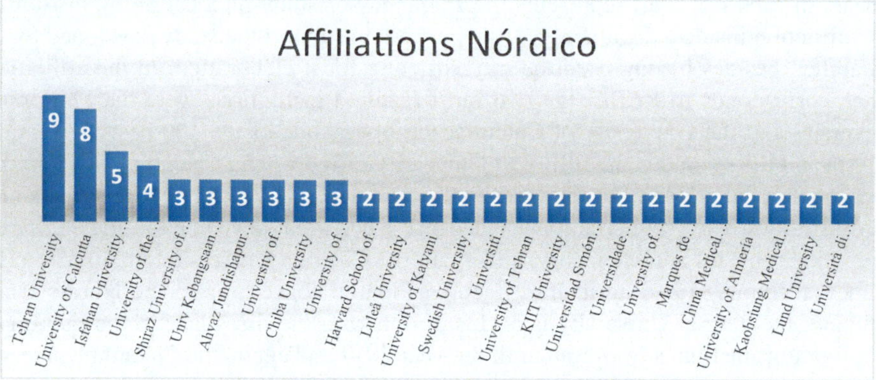

Fig. 14. Nordic questionnaire affiliations

Phase 2. Detection of the variables that determine the ergonomic methods to apply
The key topics represented on the Figs. 15, 16 and 17 are consisted to the general investigation topic, the ergonomic, applied mainly to the lower parts of the human body in footwork. It content describes the musculoskeletal disorders, the analysis of the postures

and the inconvenience to short, mid and long term that entails bad body positions while the people are working. Through the evaluation methods in the workplace, such as REBA, OWAS, Nordic Questionnaire, RULA and others, has the possibility of precise in a trustworthy way the diseases symptoms, alleviate and in the most cases avoiding that the quality life get affected, it relates mainly with occupational health.

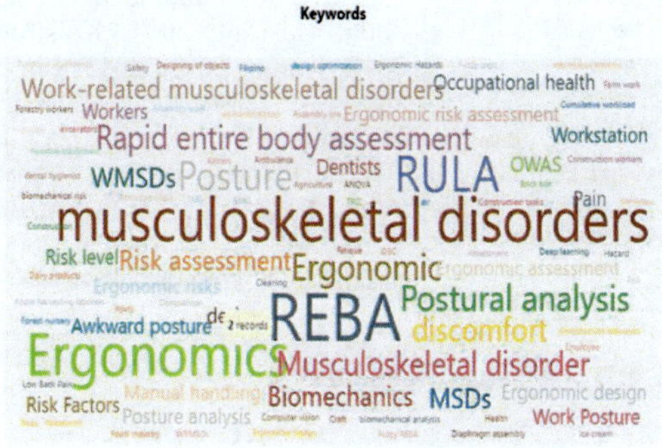

Most cited References
(Can, Atalay, & Eraslan, 2015; Chiasson, Imbeau, Aubry, & Delisle, 2012; Das, Ghosh, & Gangopadhyay, n.d.; Faaberg, Kehrli, Lager, Guo, & Han, 2010; Hignett & McAtamney, 2000; Kumar, Baliga, & Kumar, 2013; Motamedzade, Ashuri, Golmohammadi, & Mahjub, 2011; Mukhopadhyay & Srivastava, 2010; Rafeemanesh, Jafari, Kashani, & Rahimpour, 2013)

Fig. 15. Key words REBA

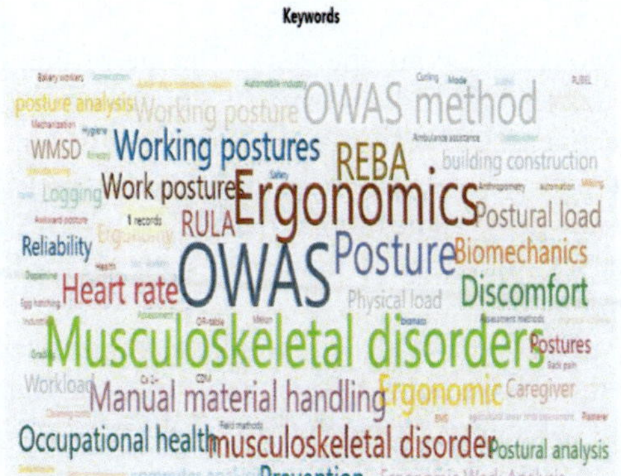

Most cited References
(Diego-Mas & Alcaide-Marzal, 2014; DOORMAAL, DRIESSEN, LANDEWEERD, & DROST, 1995; Somnath Gangopadhyay, Das, Das, & Ghoshal, 2005; Somnath Gangopadhyay, Das, Das, Ghoshal, & Ghosh, 2010; Karhu, Härkönen, Sorvali, & Vepsäläinen, 1981; Karhu, Kansi, & Kuorinka, 1977; Kee & Karwowski, 2007; Kivi & Mattila, 1991; Lee & Han, 2013; Louhevaara, 1999; Nevala-Puranen, Kallionpää, & Ojanen, 1996; Olendorf & Drury, 2001; Perkiö-Mäkelä & Hentilä, 2005; Roman-Liu, 2014; van Wendel de Joode, Burdorf, & Verspuy, 1997; Väyrynen & Könönen, 1991; Wright & Haslam, 1999)

Fig. 16. Key words OWAS

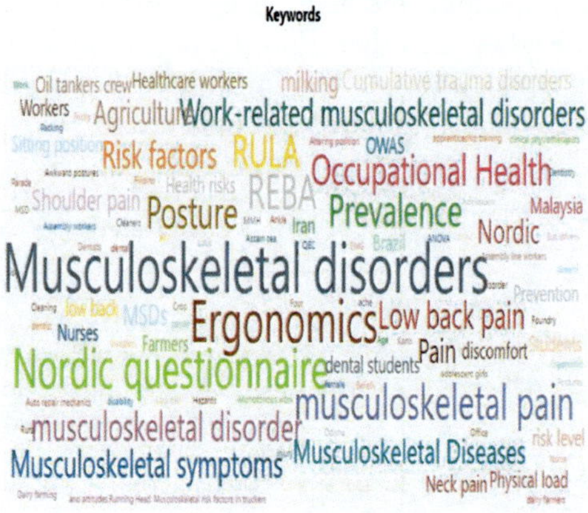	Most cited References
	(Banibrata DAS, Tirthankar GHOSH, & Somnath GANGOPADHYAY, 2013; Das & Gangopadhyay, 2015; Gallis, 2006; S. Gangopadhyay, Ghosh, Das, Ghoshal, & Das, 2010; Glover, McGregor, Sullivan, & Hague, 2005; Gummesson et al., 2006; Hussain, 2004; Kolstrup, 2012; Kumar et al., 2013; Lin et al., 2012; Mehrdad, Morshedizadeh, & Dennerlein, 2012; Palmer, 1996; Parot-Schinkel et al., 2012; Ratzon, Yaros, Mizlik, & Kanner, 2000; Singh & Chokhandre, 2015)

Fig. 17. Key words nordic questionnaire

5 Conclusions

The ergonomic has been an investigation topic arising because the interest for the creation and analysis of the workplace, next to the need of examine the body postures of the workers. This has generated the origin and inquiry of methods such as REBA, OWAS and Nordic Questionnaire, that enable the measurement of musculoskeletal affectation in all the human body extremities.

Referring to the three previous methods, which are the central theme of this bibliometric analysis, it is concluded that the search for them has had a boom from 2010 to the present, which evidences an attractiveness of research achieving that the annual average of publications in each method is increasing. In addition, the refined and selected publications of the result of the search equations indicate that the REBA, OWAS and Nordic questionnaire, are applied in the vast majority of industries such as: medical, agricultural, manufacturing, services, among others.

Also, it is identified with the geospatial results that the regions most interested in investigating this type of ergonomic methods have been the continents of Asia, Europe and America, highlighting that the most exploratory country in this category of study is India.

As for the studies in academia and affiliations, it is analyzed that the main universities with interest in this type of ergonomic research are the University of the Philippines, Technical University of Aachen and Tehran University. For the characters involved in these studies, the OWAS evaluation methods and Nordic questionnaire have the same authors in their knowledge networks, where the distinguished Indian ergonomist researcher, Gangopadhyay S, who emphasizes occupational health and the effects of postures stands out with the appearance of musculoskeletal disorders; In the REBA knowledge networks, Gangopadhyay S is also mentioned, however, it has a network with more members compared to the other methods, which is made up of Mushtaq.

F, Shah Z.A, Sheikh I.A, Ashraf M and Amjad A affiliated with the Institute of Quality and Technology Management, at the University of Punjab, Pakistan.

With the bibliometric analysis it is allowed to argue, evidence and validate the research related to the evaluation methods in a job such as REBA, OWAS and Nordic questionnaire. In this way, it could be said that an evolution in these and other ergonomic methods will arise, due to the great interest in the subject and the importance that the employee has taken in companies, since it is already necessary to analyze their well-being through Ergonomics to improve occupational health and human talent management in organizations.

References

Das, B., Ghosh, T., Gangopadhyay, S.: Child work in agriculture in West Bengal, India: assessment of musculoskele…: Discovery Service para Universidad Pontificia Bolivariana. J. Occup. Health **55**, 12–0185 (2013). https://doi.org/10.1341-9145

Can, G. F., Atalay, K. D., Eraslan, E.: Working posture analysis in fuzzy environment and ergonomic work station de…: Discovery Service para Universidad Pontificia Bolivariana. https://doi.org/10.13001884

Chiasson, M.-È., Imbeau, D., Aubry, K., Delisle, A.: Comparing the results of eight methods used to evaluate risk factors associated with musculoskeletal disorders. Int. J. Ind. Ergon. **42**(5), 478–488 (2012). https://doi.org/10.1016/j.ergon.2012.07.003

Das, B., Gangopadhyay, S.: Prevalence of musculoskeletal disorders and physiological stress among adult, male potato cultivators of West Bengal, India. Asia Pac. J. Public Health **27**(2), NP1669–NP1682 (2015). https://doi.org/10.1177/1010539511421808

Das, B., Ghosh, T., Gangopadhyay, S.: Assessment of Ergonomic and Occupational Health-Related Problems Among Fema…: Discovery Service para Universidad Pontificia Bolivariana. http://eds.b.ebscohost.com.consultaremota.upb.edu.co/eds/detail/detail?vid=11&sid=fb56bfb9-23c8-4c26-b3a5-c4c68a850577%40sdc-v-sessmgr02&bdata=Jmxhbmc9ZXMmc2l0ZT1lZHMtbGl2ZQ%3D%3D#AN=000313372100008&db=edswss

Diego-Mas, J.A., Alcaide-Marzal, J.: Using Kinect™ sensor in observational methods for assessing postures at work. Appl. Ergon. **45**(4), 976–985 (2014). https://doi.org/10.1016/j.apergo.2013.12.001

Doormaal, M.T.A.J., Driessen, A.P.A., Landeweerd, J.A., Drost, M.R.: Physical workload of ambulance assistants. Ergonomics **38**(2), 361–376 (1995). https://doi.org/10.1080/00140139508925110

Faaberg, K.S., Kehrli, M.E., Lager, K.M., Guo, B., Han, J.: In vivo growth of porcine reproductive and respiratory syndrome virus engineered nsp2 deletion mutants. Virus Res. **154**(1–2), 77–85 (2010). https://doi.org/10.1016/j.virusres.2010.07.024

Gallis, C.: Work-related prevalence of musculoskeletal symptoms among Greek forest workers. Int. J. Ind. Ergon. **36**(8), 731–736 (2006). https://doi.org/10.1016/j.ergon.2006.05.007

Gangopadhyay, S., Ghosh, T., Das, T., Ghoshal, G., Das, B.: Effect of working posture on occurrence of musculoskeletal disorders among …: Discovery Service para Universidad Pontificia Bolivariana (2010). https://doi.org/10.12107778

Gangopadhyay, S., Das, B., Das, T., Ghoshal, G.: An ergonomie study on posture- related discomfort among preadolescent agricultural workers of West Bengal, India. Int. J. Occup. Saf. Ergon. **11**(3), 315–322 (2005). https://doi.org/10.1080/10803548.2005.11076652

Gangopadhyay, Somnath, Das, B., Das, T., Ghoshal, G., Ghosh, T.: An ergonomics study on posture-related discomfort and occupational-related disorders among stonecutters of West Bengal, India. Int. J. Occup. Saf. Ergon. **16**(1), 69–79 (2010). https://doi.org/10.1080/10803548.2010.11076830

Glover, W., McGregor, A., Sullivan, C., Hague, J.: Work-related musculoskeletal disorders affecting members of the chartered society of physiotherapy. Physiotherapy **91**(3), 138–147 (2005). https://doi.org/10.1016/j.physio.2005.06.001

Gummesson, C., Isacsson, S.-O., Isacsson, A.H., Andersson, H.I., Ektor-Andersen, J., Östergren, P.-O., Hanson, B.: The transition of reported pain in different body regions – a one- year follow-up study. BMC musculoskelet. disord. **7**(1), 17 (2006). https://doi.org/10.1186/1471-2474-7-17

Hignett, S., McAtamney, L.: Rapid Entire Body Assessment (REBA). Appl. Ergon. **31**(2), 201–205 (2000). https://doi.org/10.1016/S0003-6870(99)00039-3

Hussain, T.: Musculoskeletal symptoms among truck assembly workers. Occup. Med. **54**(8), 506–512 (2004). https://doi.org/10.1093/occmed/kqh087

Karhu, O., Härkönen, R., Sorvali, P., Vepsäläinen, P.: Observing working postures in industry: examples of OWAS application. Appl. Ergon. **12**(1), 13–17 (1981). https://doi.org/10.1016/0003-6870(81)90088-0

Karhu, O., Kansi, P., Kuorinka, I.: Correcting working postures in industry: a practical method for analysis. Appl. Ergon. **8**(4), 199–201 (1977). https://doi.org/10.1016/0003-6870(77)90164-8

Kee, D., Karwowski, W.: A comparison of three observational techniques for assessing postural loads in industry. Int. J. Occup. Saf. Ergon. **13**(1), 3–14 (2007). https://doi.org/10.1080/10803548.2007.11076704

Kivi, P., Mattila, M.: Analysis and improvement of work postures in the building industry: application of the computerised OWAS method. Appl. Ergon. **22**(1), 43–48 (1991). https://doi.org/10.1016/0003-6870(91)90009-7

Kolstrup, C.: Work-related musculoskeletal discomfort of dairy farmers and employed workers. J. Occup. Med. Toxicol. **7**(1), 23 (2012). https://doi.org/10.1186/1745-6673-7-23

Kumar, S., Baliga, M., Kumar, V.: Prevalence of work-related musculoskeletal complaints among dentists in India: A national cross-sectional survey. Indian J. Dent. Res. **24**(4), 428 (2013). https://doi.org/10.4103/0970-9290.118387

Lee, T.-H., Han, C.-S.: Analysis of working postures at a construction site using the OWAS method. Int. J. Occup. Saf. Ergon. **19**(2), 245–250 (2013). https://doi.org/10.1080/10803548.2013.11076983

Lin, T.-H., Liu, Y.C., Hsieh, T.-Y., Hsiao, F.-Y., Lai, Y.-C., Chang, C.-S.: Prevalence of and risk factors for musculoskeletal complaints among Taiwanese dentists. J. Dent. Sci. **7**(1), 65–71 (2012). https://doi.org/10.1016/j.jds.2012.01.009

Louhevaara, V.: Is the physical work load equal for ageing and young blue-collar workers? Int. J. Ind. Ergon. **24**(5), 559–564 (1999). https://doi.org/10.1016/S0169-8141(98)00061-4

Mehrdad, R., Morshedizadeh, M., Dennerlein, J.T.: Musculoskeletal disorders and ergonomic hazards among Iranian physicians: Discovery Service para Universidad Pontificia Bolivariana. Archieves Iran. Med. **15**(6), 00 (2012). https://doi.org/10.1029297717353947

Motamedzade, M., Ashuri, M.R., Golmohammadi, R., Mahjub, H.: Comparison of ergonomic risk assessment outputs from rapid entire body assessment and quick exposure check in an engine oil company. J. Res. Health Sci. **11**(1), 26–32 (2011). http://www.ncbi.nlm.nih.gov/pubmed/22911944

Mukhopadhyay, P., Srivastava, S.: Ergonomic design issues in some craft sectors of Jaipur. Des. J. **13**(1), 99–124 (2010). https://doi.org/10.2752/146069210X12580336766446

Nevala-Puranen, N., Kallionpää, M., Ojanen, K.: Physical load and strain in parlor milking. Int. J. Ind. Ergon. **18**(4), 277–282 (1996). https://doi.org/10.1016/0169-8141(95)00054-2

Oborne, DJ.: Ergonomía en acción : la adaptación del medio de trabajo al hombre. Trillas (1987). https://books.google.com.co/books?id=M-VOPQAACAAJ&dq=ergonomía+en+acción,+adaptación+del+medio+de+trabajo+al+hombre&hl=es&sa=X&ved=0ahUKEwjh-a-fgbvjAhXBjFkKHdIpD3YQ6AEIKTAA

Olendorf, M.R., Drury, C.G.: Postural discomfort and perceived exertion in standardized box-holding postures. Ergonomics **44**(15), 1341–1367 (2001). https://doi.org/10.1080/00140130110085358

Palmer, K.T.: Musculoskeletal problems in the tomato growing industry: "tomato trainer's shoulder"? Occup. Med. **46**(6), 428–431 (1996). https://doi.org/10.1093/occmed/46.6.428

Parot-Schinkel, E., Descatha, A., Ha, C., Petit, A., Leclerc, A., Roquelaure, Y.: Prevalence of multisite musculoskeletal symptoms: a French cross-sectional working population-based study. BMC Musculoskelet. Disord. **13**(1), 122 (2012). https://doi.org/10.1186/1471-2474-13-122

Perkiö-Mäkelä, M., Hentilä, H.: Physical work strain of dairy farming in loose housing barns. Int. J. Ind. Ergon. **35**(1), 57–65 (2005). https://doi.org/10.1016/j.ergon.2004.08.004

Rafeemanesh, E., Jafari, Z., Kashani, F., Rahimpour, F.: A study on job postures and musculoskeletal illnesses in dentists. Int. J. Occup. Med. Environ. Health **26**(4), 615–620 (2013). https://doi.org/10.2478/s13382-013-0133-z

Ramírez Cavassa, C.: Seguridad industrial : un enfoque integral. Limusa (2000). https://books.google.es/books?hl=es&lr=&id=jDgUQb_V6PsC&oi=fnd&pg=PA5&dq=seguridad+industrial+un+enfoque+integral&ots=ZQu9A3CzmT&sig=b2J9HeODyhZCIbqNDaCgmhmoRTU#v=onepage&q=seguridadindustrialunenfoqueintegral&f=false

Ratzon, N.Z., Yaros, T., Mizlik, A., Kanner, T.: Musculoskeletal symptoms among dentists in relation to work posture: Discovery Service para Universidad Pontificia Bolivariana. Work **15**(3), 153–158 (2000). https://doi.org/10.10519815

Roman-Liu, D.: Comparison of concepts in easy-to-use methods for MSD risk assessment. Appl. Ergon. **45**(3), 420–427 (2014). https://doi.org/10.1016/j.apergo.2013.05.010

Singh, S., Chokhandre, P.: Assessing the impact of waste picking on musculoskeletal disorders among waste pickers in Mumbai, India: a cross-sectional study. BMJ Open **5**(9), e008474 (2015). https://doi.org/10.1136/bmjopen-2015-008474

de Joode, B.W.V., Burdorf, A., Verspuy, C.: Physical load in ship maintenance: hazard evaluation by means of a workplace survey. Appl. Ergon. **28**(3), 213–219 (1997). https://doi.org/10.1016/S0003-6870(96)00051-8

Väyrynen, S., Könönen, U.: Short and long-term effects of a training programme on work postures in rehabilitees: a pilot study of loggers suffering from back troubles. Int. J. Ind. Ergon. **7**(2), 103–109 (1991). https://doi.org/10.1016/0169-8141(91)90041-J

Wright, E.J., Haslam, R.A.: Manual handling risks and controls in a soft drinks distribution centre. Appl. Ergon. **30**(4), 311–318 (1999). https://doi.org/10.1016/S0003-6870(98)00036-2

Review of Applied Methodologies and Proposals Regarding Social Responsibility in the Military Forces

Angie Melissa Morales Avella[1(✉)], Anny Astrid Espitia Cubillos[1(✉)], and Jairo Andrés Agudelo Calderón[2(✉)]

[1] Industrial Engineering Program, Faculty of Engineering, Nueva Granada Military University, Bogotá, Colombia
{u2903131,anny.espitia}@unimilitar.edu.co

[2] Cavalry School, National Army of Colombia, Bogotá, Colombia
jairo.agudelo@buzonejercito.mil.co

Abstract. The promotion of actions aimed at different focuses that generally seek to contribute to a society and a better environment has been recognized as Corporate Social Responsibility. Since its inception, practical approaches to its application have been recognized in commercial, industrial, public, private organizations, among others. However, the information provided on the application of social responsibility in military institutions is scarce, specifically those belonging to Colombia, so this article will do a literature review, organization, analysis and subsequent consolidation of viable proposals for implementation in the Colombian Military Forces. After conducting the literature review, it was found that various Ministries of Defense of countries such as Spain, Chile, and the United States, among others, recognizes similar commitments in their Social Responsibility reports such as gender equality, environmental approaches that seek to reduce consumption through management system, accessibility for the disabled, prevention philosophies, family-work relationship, among others, of which (after realizing an organization and analysis of the information) three proposals are made focused on the environmental vision, the equity of gender and the humanitarian missions, that could be implemented in the Colombian Military Forces.

Keywords: Social responsibility · Military forces · Ambient · Equality · Society

1 Introduction

Social responsibility has been consolidated as a fundamental part of the institutions that seek to remain at the forefront and coexist with the context in which they develop their organizational concept, because in addition to seeking the constant financial growth of the organization. It is important to generate a contribution and have into account the social needs that impact the actors essential to their functioning, so it is relevant to the application of its fundamental aspects their recognition in the context of different types of organization.

M. T. Castañeda Galvis et al. (Eds.): ICIE 2019, LNMUINEN, pp. 102–111, 2021.
https://doi.org/10.1007/978-3-030-49370-7_11

According to the Commission of the European Communities (CCE 2001), social responsibility is defined as the voluntary decision by companies in relation to the contribution of a better society and a cleaner environment, thereby increasing competitiveness of these. Expanding this definition recognizes an integration of various aspects that have been discussed by authors such as: Méndez (2005) who specifies how social responsibility has progressed in business and how ethics and morals affect this aspect; Gómez (2006) who summarizes the relationship between social responsibility, stakeholders and the dialogue that is generated in organizations when implementing these efforts; Hernández (2011) who specifies the efficiency of social responsibility applied in five renowned corporations through a qualitative study, among others.

However, despite the fact that the information presented is wide in reference to especially commercial corporations, there is a gap in relation to military institutions that correspond to the defense of the sovereignty of a nation (Head of State 2005) as the Military forces, specifically those belonging to Colombia, for which it is proposed through a literature review the recognition of good practices found in systems similar to the object of analysis, in this case the Military forces of Colombia, where they have currently adopted strategies with this approach, with the purpose of integrating the applied approaches and reaching the consolidation of possible methodologies to be implemented as a proposal in this organization.

The research is made with the objective of creating social responsibility proposals that can be applied to the Colombian military forces.

2 Methodology

The qualitative approach of research is determined with a deductive dimension where after obtaining information, in this case by reviewing literature is possible to make an interpretation of the information, discover relationships, distribute, organize and analyze the results (Quecedo et al. 2002). Additionally the conception of interpretive compression is maintained, seeking to understand the uniqueness of the community in its context and interpret this information (Martínez 2011), in this case the application of social responsibility practices in military institutions.

The structuring of the literature review was based on that described by Gómez-Luna et al. (2014) where the process diagrammed in four steps: define the problem, search for information, organization and analysis of information, with the inclusion of a fifth step in the model that corresponds to the determination of proposals with three approaches that will be selected based on the current conditions of the institution and the conditions of the environment.

3 Results

Initially and following the scheme described in the methodology, the problem that prompted the methodological review was defined, which consists in the identification of the main approaches and good practices carried out worldwide around social responsibility for military institutions with similarities to the Colombian Military Forces composed of the National Army, the National Navy and the Air Force.

Having clear the problem, a process of locating documents from various sources such as web pages, research articles, databases of international journals, among others, was selected, selecting the relevant information according to the context and according to certain criteria such as the actuality of the information, defining a maximum range of five years prior to the current date, seeking the novelty of the information. Also in this step, variations in the treatment of information were recognized, according to each institution since some institutions treat it directly as a Social Responsibility Policy and others as the philosophy of the organization or impact on society. Subsequently, it was analyzed. Table 1 summarizes the main common thematic axes applied in relation to social responsibility with the country and the institution to which its application is attributed.

Table 1. Thematic axes Social Responsibility Military institutions

Social Responsibility in Military Institutions				
Country	Institution	Main theme	Application	Sources
France	French Armed Forces	Civic action	Military social security, center for disabled people, support for the injured and family, education, social development, training access programs	GR20 Course Conference. Social action in the French FAS. TCOL Bertrand
Chile	Chilean Armed Forces	Equity, Diversity and Inclusion	The institutional policy seeks to include every person who wants to serve Chile. Incorporation of people with disabilities	2017 Army Sustainability Report. Book of the National Defense of Chile 2017
			Participation of women, promulgation of modification to the "Escalafón of the Women's Military Service". Inclusion of female personnel in Peace Operations	
			Command Order that dictates the procedures to be followed in the face of workplace and sexual harassment	
		Civic and Humanitarian Action	Recognition of the social, cultural, religious or spiritual values and practices of indigenous peoples, supporting their development and seeking to generate spaces for participation and cooperation	

(*continued*)

Table 1. (*continued*)

Social Responsibility in Military Institutions				
Country	Institution	Main theme	Application	Sources
			Promulgation of Social Responsibility Policy, initiatives related to the environment, biodiversity, support against natural disasters, the promotion of the traditions and the protection of cultural historical heritage, health care for citizens, and through the work of sovereignty and cohesion social in remote areas of the national territory. Transparency, development cooperation	
U.S	United States Army	Equity, Diversity and Inclusion	They have a sexual harassment prevention and response program which organizes training sessions to teach how harassment can be prevented or what to do in case it is happening. They also provide a support network	Web Army of the United States
	United States Air Force	Community	They have necessary supplies so that the community, both military and civil (wives of military and children), has all the facilities such as schools, supermarkets, police stations and firefighters. A small citadel where civil participation is taken into account for its operation	Web Army of the United States
UK	Royal Army	Civic and humanitarian action	Rapid intervention in disasters such as evacuated due to flood risks, disposition of the people, implements and collaboration	British Royal Army website
			It has programs for the prevention of mental illnesses. Also, they have a welfare service for both members and their families, providing assistance of various kinds	

(*continued*)

Table 1. (*continued*)

Social Responsibility in Military Institutions				
Country	Institution	Main theme	Application	Sources
	Royal Navy	Equity, Diversity and Inclusion	It has a policy of gender equality, which states that everyone has the same s opportunities, is also a pleasant working environment	British Royal Navy website
		Civic action	It has a welfare service for both members and their families, offering different kind assistance is with personnel trained constantly faces problems that corps members and their families may face	
Sweden	Swedish Armed Forces	Environment	Due to the importance of reducing the environmental impact of the country, this objective is implemented within all the country's military forces, making it possible for its operation to have the least negative environmental impact	Swedish military forces website
Spain	Spanish Armed Forces	Multidisciplinary	All member institutions have been implementing various policies aimed at l to social responsibility with its different categories: environment, gender equality, human rights, transparency, boost the industry, preservation of historical and cultural heritage, among others, with an agenda quite complete and comprehensive that fully meets all aspects of social responsibility	Social Responsibility Report. Ministry of Defence
Mexico	Mexican Army and Air Force	Civic Action	Search for social welfare of the population in marginalized or extreme poverty areas. Health campaigns, sports promotion and reading rooms, transportation and distribution of free textbooks. Support Force for disasters	Web of the Government of Mexico

(*continued*)

Table 1. *(continued)*

Social Responsibility in Military Institutions				
Country	Institution	Main theme	Application	Sources
Ecuador	Armed Forces of Ecuador	Civic and humanitarian action	Carrying out activities of support for the development of state institutions. Deployment of humanitarian aid, technical visits, humanitarian demining. Activities in relation to human rights. Emergency supports	Bulletin of Armored Cavalry N ° 109. Management Report of the Ecuadorian Army 2018
		Equity, Diversity and Inclusion	Activities in relation to human rights, gender equity and international law	
		Environment	Agreement with Pro Forestry for the production of plants in nurseries, savings and efficiency in the levels of resource consumption, reduction of waste generation	
Canada	Canadian Armed Forces	Transparency	Implementation of a value-based defense ethics program implemented to encourage the practice of ethics in the workplace and in operations	Web of the Government of Canada
		Environment	The Canadian armed forces intervene in the report that plans to address global warming, natural resources and the conflicts they cause	
		Equity, Diversity and Inclusion	The Canadian Forces are world leaders in military gender integration, having a very high proportion of women in the forces and administrative areas	Canada Armed Forces website
			They have educational programs and income plans for indigenous peoples	
			In partnership with the Center for Response to Sexual Behavior, the contribution program of the sexual assault center is launched to respond to gender-based violence	

Taking into account the recognition of the thematic axes in order to distinguish the approach to be considered for the final phase or consolidation of the proposal, a statistical analysis is carried out that includes the frequency of each axis in the total population analyzed. These results are represented in Diagram 1.

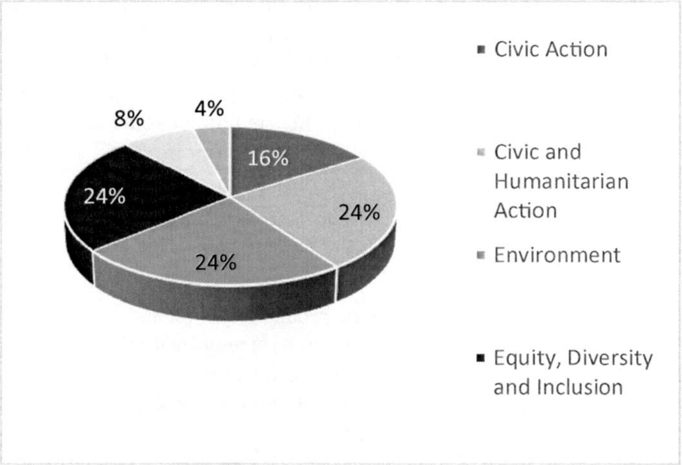

Diagram. 1. Circular diagram thematic axes of Social Responsibility.

Analyzing Diagram 1, it is possible to recognize the strong participation of the thematic axes of civic and humanitarian action, diversity and inclusion, and the environment and equity, for which reason it is decided to focus the proposals on these three themes. This selection was endorsed trough the expert judgment, which determined selected topics have a breadth and feasibility enough for application in the Military Forces of Colombia, making possible the selection and consolidation of the proposals based on recognized best practices around thematic.

It should be noted that the work of the Colombian Military Forces in the field of social responsibility is quite extensive and is constantly updated and growing. For example, in terms of equity, diversity and inclusion, military institutions have made reforms in such a way that there is participation of both women and men in all the functions and tasks of the institutions, allowing women to access any position within the structure of the organization. Similarly, the military forces advance civic and humanitarian actions, an example of this is the Binational Civic Action Day between Colombia and Ecuador, where the intention is to offer health care of different specialties to the inhabitants of the border areas. On the other hand, in relation to environmental responsibility, actions such as the extinction of forest fires are carried out with the intention of preventing them from spreading and that there is no serious and irreparable environmental damage.

Therefore, the proposals seek to mention those aspects in which the armed forces can improve their social responsibility by making it increasingly complete and comprehensive, without ignoring existing advances, namely: Equity, Diversity and Inclusion; Civic and Humanitarian Action; and Environment.

4 Discussion

In this way, the first proposal is related to the categories: Equity, Diversity and Inclusion, where, specifically, it focuses on the incorporation of the indigenous population and people with reduced abilities within the military forces based on the Sustainability Report 2017 (Army Personnel Direction 2017) of the Chilean Army in which they describes how they have taken into account the indigenous populations both for respecting their values and traditions and for their presence within the members of the body reaching agreements that allow army personnel to be trained in Mapuche and Aymara dialects, culture and worldview. Similarly, but with a focus on the administrative structure of the institution, the Chilean army has the initiative of incorporating people with reduced capacities into its workforce, as well as supporting foundations that benefit this population and scholarships that can access higher education.

The second proposal is focused on the category of Civic and Humanitarian Action, which takes into account the role of military forces to protect historical heritage and assets of cultural interest, considering this and taking as a reference the Army of Peru (Borda 2019), Colombian military forces could generate multidisciplinary commissions for the archaeological exploration of the country in search of protecting the existing heritage and the one that has not yet been discovered, taking advantage of the technical and tactical capabilities of the military institutions and the capacity with which It is counted to reach any territory of the country.

The third proposal focuses on the Environment category, it is suggested the reduction of energy consumption with the improvement of staff facilities, considering the installation of solar panels for sanitary water heating, instead of using fossil fuel as gas for water heaters, or where appropriate the presence of heaters since there may be an absence of these. In addition to the installation of solar panels and other types of environmentally friendly power generation sources, timers can be installed in water heaters so that their use is controlled and efficient. Similarly, it can be sought that buildings have designs that take full advantage of natural lighting, so that during the day there is no need to use electricity or other energy for artificial lighting. All this would go hand in hand with the necessary trainings to the staff in order to appropriate the knowledge of energy efficiency, so that their behavior is consistent with the changes to be made (Army Personnel Directorate 2017).

Finally, as expected results after the application of these proposals, it is expected to generate synergies between the interested parties and to arrive at projections based on recognized functional innovations regarding social responsibility, according to the international context.

5 Conclusions

Since its emergence, social responsibility has influenced the organizational policies of the various institutions that seek to be part of a collaborative and more competitive environment where the objectives go beyond financial values and come to be considered visions that take society into account as an axis. The military forces are not the exception, therefore, in this article a bibliographic review of documents referring to documented experiences of social responsibility in the Military forces was carried out so that, through the different organizational and analysis strategies, it was possible to recognize wide fundamental points approximations other institutions on this subject, recognizing as such to: 1) humanitarian and civic action, 2) fairness, diversity and inclusion and 3) environment.

The recognition of these fundamental axes and good practices in other military institutions worldwide provided the tools to propose similar methodologies applicable in Colombian military forces to strengthen and complement the actions undertaken in this regard.

The first proposal focused on equity, diversity and inclusion seeks to consider the incorporation of indigenous population and people with reduced skills within the forces, adding to this respect for their culture, the promotion of training in related topics and support for foundations that they benefit this population.

The second proposal within the category Civic and humanitarian action proposes the generation of multidisciplinary commissions for the protection of heritage, taking advantage of the capabilities and scope of the organization.

And the third proposal focuses on the care of the environment looking for a change in the habits of energy consumption and taking into account financial analyzes that support the decisions the possible installation of systems that avoid the use of fossil fuels, considering in a transversal way to the process training of interested parties.

Acknowledgements. Product derived from the development of the research project INV-ING-2987 entitled "Identification of the technological innovations required by the Colombian armed forces in the context of the post- agreement as part of the fulfillment of their Social Responsibility from the theory of stakeholders and the common good" funded by the Office of the Vice President for Research of the Military University Nueva Granada, effective 2019–2020.

References

Army Personnel Direction: Army Sustainability Reports. Military Geographic Institute, Santiago de Chile (2017)

Borda, J.: Army Press: Peruvian Army. Obtained from the Peruvian Army, 21 July 2019. www.ejercito.mil.pe/index.php/prensa-ejercito/noticias/item/1691-ejercito-se-suma-al-gran-proyecto-turistico-e-historico-en-el-sitio-arqueologico-el-gran-pajaten

Commission of the European Communities (CC3): Promote a European framework for corporate social responsibility. Brueslas, 366 (2001). http://www.europarl.europa.eu/meetdocs/committees/deve/20020122/com(2001)366_es.pdf

Gómez, V.: Corporate social responsibility and social dialogue" at the International CSR Conference. Carolina Foundation and América Foundation. ISSN-e 1885.9119, No. 7, pp. 13–16 (2006). https://dialnet.unirioja.es/servlet/articulo?codigo=2857678

Gómez-Luna, E., Fernando-Navas, D., Aponte-Mayor, G., Betancourt-Buitrago, L.: Methodology for bibliographic review and information management of scientific topics, through its structuring and systematization. Dyna **81**(184), 158–163 (2014). http://www.redalyc.org/articulo.oa?id=49630405022

Head of State: Organic Law 5/2005, on National Defense. State Agency Official State Gazette, Section I, general provisions. Recovered from: BOE-A-2005-18933 (2005)

Hernández, C.: Reality of corporate social responsibility. Miguel Hernández Communication Journal. ISSN-e 1989-8661, No. 2, pp. 82–92. Miguel Hernández University, UMH (2011). http://mhcj.es/2011/05/25/cirohernandez

Martínez, J.: Qualitative Research Methods. Magazine of the International Corporation for Educational Development, vol. 8 (2011)

Méndez, M.: Ethics and corporate social responsibility at ICE: Economy magazine. ISSN 0019-977X, No. 823, pp. 141–150 (2005). https://www.researchgate.net/publication/28096050_Etica_y_responsabilidad_social_corporativa

Quecedo, R., Castaño, C.: Introduction to qualitative research methodology. J. Psychodidactics (14), 5–39 (2002). http://www.redalyc.org/articulo.oa?id=17501402

Sales Forecasting Difficulties' Analysis on Colombian Direct Sales Companies

Carlos A. Castro-Zuluaga(✉) and Mariana Arboleda(✉)

Production Management and Logistics Research Group, Faculty of Engineering, Universidad EAFIT, Medellín, Colombia
{ccastro,marbol12}@eafit.edu.co

Abstract. Direct Selling Industry reports billions on sales in Colombia and all around the globe. It is unique among retail channels because instead of relying on traditional marketplaces, direct selling companies maintain salesforces of millions of independent consultants who make direct contact with each of their clients. Direct selling companies, like most others, need to make good projections of their demand at the stock keeping unit level to properly manage their inventory. However, the literature has ignored them, so as of today there are no solutions focused on the singularity of the business. This means companies are forecasting through empirical methods. Our ongoing research seeks to recognize the main obstacles that the sector faces and to find strategies that have been proposed for other industries and could be adapted to improve prediction accuracy.

Keywords: Direct sales · Sales · Forecasting · Practices

1 Introduction

According to statistics from the World Federation of Direct Selling Associations (WFDSA), direct selling is a $ 193 billion annual business, consisting of 118.4 million independent representatives. It has maintained a constant growth over time, with a 1.7% annual rate between 2015 and 2018 (Wfdsa.org 2019a). In Colombia, for the last three years the industry has experienced a similar growth rate, reporting sales of $ 2.4 billion dollars (approximately $ 7350 billion Colombian pesos, according to the exchange rate at the closing date of the report), with 2,276,225 independent affiliated representatives (Wfdsa.org 2019b); of which 89% are women and 85% live in neighborhoods of low and medium-low socioeconomic status (Kühn et al. 2017). All this indicates that the industry is not only profitable but that, due to its characteristics in terms of independence, flexibility, and level of reward (Seldia 2018), it strongly impacts the economy of the countries in which it has a presence and is projected to remain in the future.

Direct selling is then understood as any "business model in which a company uses person-to-person interactions to sell its products" (O' Connell 2019), normally, product promotion is done through printed catalogs that the representatives send their clients

M. T. Castañeda Galvis et al. (Eds.): ICIE 2019, LNMUINEN, pp. 112–118, 2021.
https://doi.org/10.1007/978-3-030-49370-7_12

and thus generate sales in their area of influence, without having a structured store or making, in most cases, a capital investment.

There are records of direct selling associations and companies since 1910 (Dsa.org 2019), but retail literature seems to completely ignore them (Ridgway and Peterson 1989), so there is little information about guidelines and strategies to optimize their supply chain. There is evidence about how traditional and empirical methods are still used to forecast future sales. For example, in the World of Direct Selling web page, sales forecasting is seen and explained as a cross-functional exercise that salesforces must "understand and embrace" (Murphy 2018). Another article mentions that they see planning as "a mixture of science and informed intuition" (Wds 2016); contrary to the demands of the current market and industry 4.0 that prioritize the intelligent use of data and predictive models to replace traditional models (Forger et al. 2017).

Direct selling demand is more complicated and has greater variability compared to many other retailers. Catalogs must be perceived as unique for each sales cycle so they offer at least 20% of new and fashionable products, intermittent discounts, and a whole new layout with different levels of product exposure. Therefore, subjecting the products to constant structural changes. Moreover, there is a large number of independent variables with unknown effects such as the variability of salesforce and how they distribute geographically.

As far as we know, only two case studies have been carried out in Latin America to propose automated methods for demand forecasting in direct sales companies. One of them created a linear regression model. It aims to predict the total income of the next sales cycle restricted to the largest sales region and age group of sellers. It does not mention how they scale the numbers to the stock-keeping unit (SKU) level and assumes a constant effect of discounts and other variables (Sandoval 2017). The second study, in Venezuela, discusses the creation of 52 multivariate regression models for the forecasting of 26 different product families. They integrate 31 continuous and categorical variables to describe each product (Millan and Boada 2010; Boada et al. 2012a, b). However, their methodology is impossible to replicate.

2 Methodology

The main objective of this paper is to identify the most common challenges direct sales companies face when forecasting their demand, in order to provide a useful guide to improve the forecasting process and its accuracy.

(i) Interviews: We conducted semi-structured interviews with two different companies about their forecasting processes, their performance, expectations, decision process and the reasons they believe are impacting their results. We were also able to observe five analysts while doing their job. Therefore we acquired an important understanding of the input data they receive, the timeframes they work with and the decisions they make when faced with uncertainty. No specific questions were asked about their products, prices, or their employees identities. Results were collected and generalized by the investigators in such a way that companies are not identifiable in a country with over 20 recognized direct selling associations.

(ii) Literature review: We retrieved information about the identified challenges in specialized databases and journals through three main topics: fast fashion forecasting (for intermittent and new product demand), forecasting in the presence of structural change, and best practices in business forecasting.
(iii) Real data analysis: We also had the opportunity to access the data of a direct sales company and were able to identify difficulties inherent to how they collect, perceive and utilize their data.

3 Results

(i) Companies agreed that the biggest difficulties they face are directly related to how they conceive their catalogs. They know that the sense of novelty of every catalog pushes their clients to buy on each campaign, but also adds variation to their demand. Every catalog offers around 4000 different SKU, most of them are intermittent between campaigns (to generate a sense of urgency in the clients), so demand history is sparse. Even when a product is on two or more consecutive campaigns, it is exhibited in different positions throughout the catalog; and just as in a retail shop the top shelf gets more attention than the lower shelf, companies have perceived that there are some pages that get more attention than others. Also, selling cycles differ from seasons and traditionally-seen calendars. Companies define special offers and targets according to their own marketing strategies. For example, it is a common practice to have special discounts for their companies' anniversaries which is fixed in a specific campaign, not a date; or to focus one campaign on beauty products and the next one on one-fits-all family products.

Another challenge they agreed on is that they must present forecasts three to four months in advance, even before the catalog's final design is approved, which means lower responding rates to changing trends and even more uncertainty about the effect some variables will have. On the other hand, companies acknowledge that their performance is dependent on their salesforce. Their geographical distribution, their motives, and how they relate to the company can improve or restrict a whole campaign results. Companies offer incentives and promote loyalty between their representatives, and witness their impact. However, they cannot yet predict exactly how good or bad will an incentive be perceived and how many new representatives and clients it will, in fact, attract.

Figure 1 Factor segmentation summarizes the results in a two-dimensional space. The horizontal axis denotes whether the company can decide over the factors and the vertical axis denotes whether there is historical information collected about the factors. This way, we can classify and make decisions about how we are going to use them in our research. For instance, companies cannot directly set the values for the variables classified in the first quadrant, but they collect plenty of information that can be studied to gain insight into the business, get to measure their impact on sales and later use them as parameters in a forecasting model. Variables on the second quadrant can be as well studied, measured and be used on experimental analysis to test different hypotheses or perhaps improve future forecasting accuracy.

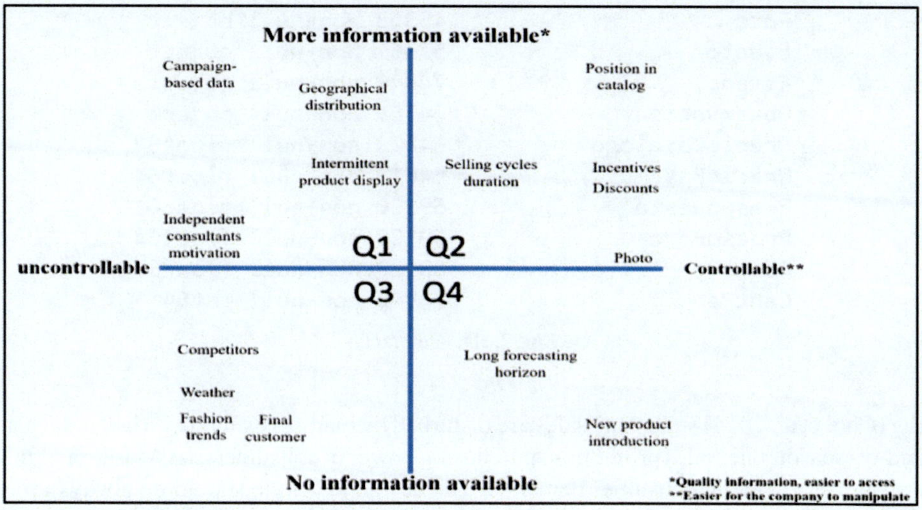

Fig. 1. Factor segmentation

Variables on the third and fourth quadrant add complexity to the demand and companies should take actions to reduce their impact on forecasting and probably set accuracy expectations accordingly (Gilliland and Sglavo 2010).

(ii) Challenges stated above have been studied separately in different retail sectors as **demand uncertainty and unknown demand distribution** (Huber et al. 2019; Kilimci et al. 2019), **highly volatile data patterns, high variety of products and big dimensionality** (Wong and Guo 2010; Choi et al. 2014) **continuous structural change and variable impact of marketing strategies** (Huang et al. 2019) and **intermittent and new product demand forecasting** (Fundamentals 2016; Vandeput 2019).

(iii) During our first exploratory analysis, we found that the data available is highly unstructured. It presents missing data, mixed information and many descriptive fields about marketing strategies. There are also some variables that are not yet being measured, and many others that are not strictly differentiated. We also noticed that some notation has changed over time and historical information has not been updated. This means that thoughtful data cleaning and preprocessing is going to be necessary before implementing any methodology, and that, as usual, companies need to define better data acquisition practices and train their workers to achieve better performance Fig. 2 Data extract is an example of one dataset and its disparity between columns.

4 Discussion

The aforementioned studies propose methodologies to cope with the sales forecasting constraints they found. However different their sector of application is, there are important similarities that lead us to adapt and incorporate their findings in our own.

```
Foto                    46363 non-null object
Evento                  57367 non-null object
Estado                  74840 non-null object
Observacion             74769 non-null object
PrecioCatalogo          64800 non-null float64
PrecioBase              64813 non-null float64
Presupuesto             69120 non-null float64
PrecioPpto              69120 non-null float64
CostoOperativo          69120 non-null float64
CantPed                 74840 non-null int64
```

Fig. 2. Data extract

E.g (Choi et al. 2014) state that "sales are disturbed by many exogenous variables such as end-of- season sale, sales promotion, purchasing power of consumers, (...) fashion trends provide very volatile consumer demands. The design and style should be always up to date and most of the items are not renewed for the next collection. Consequently, historical sales are often not available since most of the items are **ephemeral.** Product variety is huge. Indeed items are declined in many color alternatives to meet the fashion trend, and in various sizes which should match with morphologies of the target consumers" and continue to explore different approaches that have been successfully implemented, including a three-step process of clustering historical products that have similar sales profile and assigning each new product to one sales profile from its descriptive criteria.

(Barrow and Kourentzes 2016) concluded "that forecast combinations improve forecast accuracy and bias, in agreement with the literature, but also result in more normally distributed errors" and that those combinations have beneficial implications for inventory management. They also agreed with (Green and Armstrong 2015) in their review Simple versus complex forecasting, when they realized that "simple [models], such as the Median, performed at least as good, if not better, than more complex methods", and with (Huber et al. 2019) that found that "more sophisticated estimation methods such as ANNs and Gradient Boosted Decision Trees require more training data in order to produce reliable results" and as we concluded before, having that amount of data about every single SKU is not always possible.

Simplicity over complexity in forecasting may sound counterintuitive, but it is also what (Gilliland and Sglavo 2010; Gilliland 2015) discuss in their work. They affirm that simpler approaches (even as simple as näive) are most of the time more accurate than long decision processes. In some cases, companies would feel so overwhelmed about their amount of information and responsibility that would skip the simpler methodologies to make judgmental overrides. This practice leaves companies with no baseline forecast to compare to, and possibly unreachable expectations, therefore it is important to evaluate the performance of each forecasting step, or added complexity through the Forecast Value Added Analysis.

Future work will include proposing a methodology to forecast sales in a direct sales company guided by our findings and measured step by step of increased complexity to verify that value is added to the forecasting process.

References

Barrow, D.K., Kourentzes, N.: Distributions of forecasting errors of forecast combinations: implications for inventory management. Int. J. Prod. Econ. **177**, 24–33 (2016). https://doi.org/10.1016/j.ijpe.2016.03.017

Boada, A.: Diseño de un Modelo de Predicción de Demanda, ajustado a empresas de estilo de venta directa. Memorias Arbitradas en VIII Congreso de Investigación y Creación Intelectual de la UNIMET (2012). https://www.researchgate.net/publication/282292032_Diseno_de_un_Modelo_de_Prediccion_de_Demanda_ajustado_a_empresas_de_estilo_de_venta_directa

Boada, A.: Modelo lineal dinámico bayesiano como procedimiento de actualización automática para modelos estadísticos predictivos. Dimens. Empres. **15**(1), 33–52 (2017). https://doi.org/10.15665/rde.v15i1.547

Choi, T.M., Hui, C.L. Yu, Y.: Intelligent Fashion Forecasting Systems: Models and Applications, Intelligent Fashion Forecasting Systems: Models and Applications (2014). https://doi.org/10.1007/978-3-642-39869-8

Dsa.org: Who we are (2019). https://www.dsa.org/about/association. Accessed 9 Sept 2019

Forger, G., et al.: Material handling and logistics US. Roadmap 2.0. Charlotte (2017). http://www.supplychain247.com/paper/material_handling_logistics_us_roadmap_2_0/apics#register

Fundamentals, L.: CTL.SC1x - supply chain & logistics fundamentals forecasting for special cases. MIT Center for Transportation and Logistics (2016)

Gilliland, M., Sglavo, U.: Focus on forecasting: worst practices in business forecasting. Analytics (2010). http://analytics-magazine.org/focus-on-forecasting-worst-practices-in-business-forecasting/

Green, K.C., Armstrong, J.S.: Simple versus complex forecasting: the evidence, March 2015

Huang, T., Fildes, R., Soopramanien, D.: Forecasting retailer product sales in the presence of structural change. Eur. J. Oper. Res. (2019). https://doi.org/10.1016/j.ejor.2019.06.011

Huber, J., et al.: A data-driven newsvendor problem: from data to decision. Eur. J. Oper. Res. **278**(3), 904–915 (2019). https://doi.org/10.1016/j.ejor.2019.04.043

Kilimci, Z.H., et al.: An improved demand forecasting model using deep learning approach and proposed decision integration strategy for supply chain. Complexity (2019). https://doi.org/10.1155/2019/9067367

Kühn, S., et al.: Resumen ejecutivo, World Employment and Social Outlook (2017). https://doi.org/10.1002/wow3.98

Gilliland, M.: Forecast value added analysis: step by step. Product Marketing, pp. 1–27 (2015). https://www.sas.com/content/dam/SAS/en_us/doc/whitepaper1/forecast-value-added-analysis-106186.pdf

Millan, A., Boada, A.: Predicción de la demanda de productos en empresas de venta directa – aplicación de regresión múltiple y series temporales en la psicología del consumo. Trabajo desarrollado en co-autoría con el profesor: Antonio Boada (USB). Caracas: Venezuela (2010). https://www.researchgate.net/publication/215689034_Prediccion_de_la_demanda_de_productos_en_empresas_de_venta_directa_-_aplicacion_de_regresion_multiple_y_series_temporales_en_la_psicologia_del_consumo_Trabajo_desarrollado_en_co-autoria_con_el_profeso

Murphy, D.: The right way to create a sales forecast in direct selling, World of Direct Selling (2018). https://www.worldofdirectselling.com/right-way-to-sales-forecast-in-direct-selling/. Accessed 21 July 2019

O' Connell, L.: Global direct selling market - statistics & facts—statista, statista.com (2019). https://www.statista.com/topics/4883/direct-selling-market/. Accessed 22 July 2019

Ridgway, N.M., Peterson, R.A.: Consumers who buy from direct sales companies information seeking on competitors and perception of competition by food store managers view project TYPE IV error in marketing research: the investigation of ANOVA interactions view project. J. Retail. **65**(2), 273–286 (1989). https://www.researchgate.net/publication/236965116. Accessed 10 July 2019

Sandoval, J.C.: Estimación de Ventas, en una Industria Cosmética de Venta Directa mediante Minería de Datos. Universidad Técnica Federico Santa María, pp. 1–20 (2017)

Seldia: Direct selling in Europe: 2018 retail sales, Brussels (2018). https://www.seldia.eu/images/pdf/Statistics_Overview_countries_2019.pdf

Vandeput, N.: Croston forecast model for intermittent demand, Medium (2019). https://medium.com/analytics-vidhya/croston-forecast-model-for-intermittent-demand-360287a17f5f. Accessed 7 Sept 2019

Wds: The route to better sales forecasting. The World of Direct Selling (2016). https://www.worldofdirectselling.com/route-to-better-forecasting/. Accessed 9 Sept 2019

Wfdsa.org: 2018/2019 annual report WFDSA, Washington (2019a). https://wfdsa.org/download/advocacy/annual_report/WFDSA-Annual-Report-2019.pdf/?lang=es. Accessed 22 July 2019

Wfdsa.org: Global direct selling - 2018 retail sales, Washington (2019b). https://wfdsa.org/wp-content/uploads/2019/06/Sales-Seller-Report-FINAL.pdf. Accessed 21 July 2019

Wong, W.K., Guo, Z.X.: A hybrid intelligent model for medium-term sales forecasting in fashion retail supply chains using extreme learning machine and harmony search algorithm. Int. J. Prod. Econ. **128**(2), 614–624 (2010). https://doi.org/10.1016/j.ijpe.2010.07.008

Study to the Work Station in a Gas Station

Silvia D. Arciniegas(✉), Maria M. Muñoz, Marlon S. Ramírez(✉), Jairo Nuñez, and María del Coral Pérez Ordoñez(✉)

Facultad de Ingeniería Industrial, Escuela de Ingenierías, Universidad Pontificia Bolivariana – Seccional Bucaramanga, Santander, Colombia

{silvia.arciniegas.2016,maria.munoz.2016,marlon.ramirez.2016, jairo.nunez,maria.perezor}@upb.edu.co

Abstract. Nowadays ergonomics is very important to study the relationship that the employee has with the workplace, because it depends on a healthy environment avoiding discomfort and diseases, mainly generated by incorrect posture, affecting the employee's quality of life and organizational productivity. Thus, this research aims to analyze the different aspects that influence working conditions by applying an ergonomic checklist of 128 factors grouped into 10 areas to a gas station, in conjunction with the use and development of Rapid Entire Body Assessment (REBA) methods, Working Posture Analysis System (OWAS) and the Nordic questionnaire, these being observational procedures on the different positions adopted by the worker, in order to select practical solutions to problems or specific risk agents becoming corrective actions that help improve the workplace.

Keywords: Ergonomics · Nordic questionnaire · OWAS · LCE · DMEs · REBA

1 Introduction

The interest in adapting things to human beings has grown progressively over the years in all the tasks or positions that exist, combining elementary disciplines such as psychology, engineering and physiology, which help to prevent health problems that affect the workers' quality of life. In order to consider an ergonomic study in a gas station (GS) of liquid hydrocarbons, it is necessary to evaluate different aspects of the design of the workplaces of the employees who provide the tanker service, called gas station attendant, since they are directly responsible for the loyalty of the clients with the gas station; directly influencing the productivity of the same. In order to guarantee the comfort, safety and health of the worker in the work area, it is necessary to identify the risks that exist in the work environment, to observe their positions in order to determine the possible movements that generate musculoskeletal disorders in the person and that continuously affect performance in a graduated scale, avoiding the achievement of objectives that could generate success. Therefore, in the ergonomics and study of workplaces, methods were created to evaluate the level of risk, body involvement and the most affected area in the worker's body. On the other hand, the Rapid Entire Body Assessment (REBA), Working Posture Analysis System (OWAS), Check List and Nordic Questionnaire procedures were applied in the GS since they allow checking the upper and lower extremities of the gas station attendants in charge of handling fuel dispensers.

M. T. Castañeda Galvis et al. (Eds.): ICIE 2019, LNMUINEN, pp. 119–128, 2021.
https://doi.org/10.1007/978-3-030-49370-7_13

2 Objectives

Assess risk factors that could cause work-related musculoskeletal disorders (MSDs) to gas station employees.

- Identify risk factors in different body postures.
- Recognize the body zones with the greatest musculoskeletal affectation in the work of the gas station attendant.
- Propose adequate preventive measures to improve working conditions at the service station in accordance with the assessment of the physical load derived from the postures adopted during work.

3 Methodology

This research is a comparison as it allows an analytical and systematic procedure to contrast four ergonomic assessment methods (Nordic Questionnaire, OWAS, REBA and Check List) applied to two GS operational workers, gas station attendant, one in charge of the motorcycle fuel island or dispenser and the other one for automobiles.

In the first phase, a global evaluation of the ergonomics of the GS was carried out through the ergonomic checklist of 128 factors whose objective is to map the initial identification of occupational risks in the organization, leading to possible preventive actions and recommendations in each factor as long as it is not being complied with.. To this end, 10 work areas or verification points adapted to the GS were determined, such as the handling and storage of materials, manual tools, safety of production machinery, workplace design, lighting, facilities, environmental risks, hygienic services and rest areas, personal protective equipment and work organization. ("LEST Method - Global Job Evaluation" 2015.)

Subsequently, the Nordic questionnaire is carried out, in which the level of occupational risk was estimated by identifying each of the symptoms and musculoskeletal affectations in the gas station attendant, in order to recognize which of the body parts is most affected in the performance of their work tasks. In addition, the postural risk is evaluated in terms of frequency by gravity, carrying out a simple and systematic classification of the work postures, together with observations on the development of the tasks. This is known as the OWAS method, in which possible combinations of back, arm and leg positions and the magnitude of the load that is manipulated while adopting the posture are identified (Fig. 1). In this way, it gives way to the configuration of the "posture code" with the four digits assigned to each observed posture according to the previously mentioned positions and the fifth digit which determines the phase in which the coded posture has been observed (Table 1), all with the aim of obtaining an assessment of the discomfort or risk that determines the category of risk to each part of the body according to the relative frequency of the various positions adopted in the observed postures. (Table 2) (annex 1: owas methodology, n.d.; owas method for evaluating working positions – introduction 2019.)

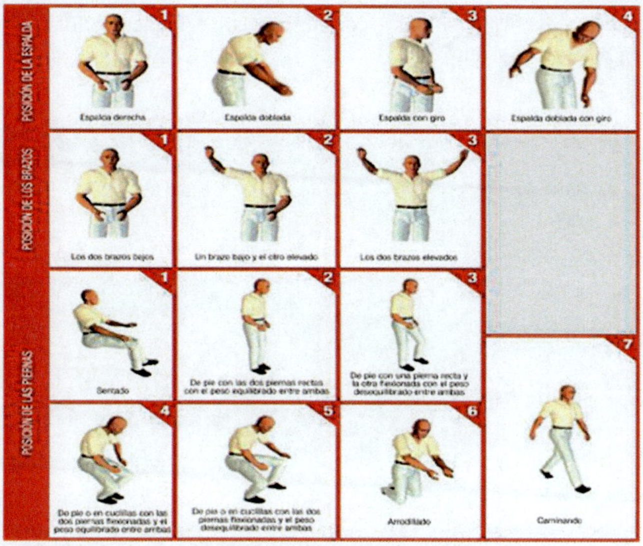

Fig. 1. Back, arms and legs postures. Source ("HSEC Magazine - OWAS: A Practical and Simple Method of Ergonomic Assessment" 2013)

Table 1. Risk category by posture code. Source ("OWAS Method - Ovako Working Analysis System" 2015)

	Legs	1			2			3			4			5			6			7		
	Load	1	2	3	1	2	3	1	2	3	1	2	3	1	2	3	1	2	3	1	2	3
Back	Arms																					
1	1	1	1	1	1	1	1	1	1	1	2	2	2	2	2	2	1	1	1	1	1	1
	2	1	1	1	1	1	1	1	1	1	2	2	2	2	2	2	1	1	1	1	1	1
	3	1	1	1	1	1	1	1	1	1	2	2	3	2	2	3	1	1	1	1	1	2
2	1	2	2	3	2	2	3	2	2	3	3	3	3	3	3	3	2	2	2	2	3	3
	2	2	2	3	2	2	3	2	3	3	3	4	4	3	4	3	3	3	4	2	3	4
	3	3	3	4	2	2	3	3	3	3	3	4	4	4	4	4	4	4	4	2	3	4
3	1	1	1	1	1	1	1	1	1	2	3	3	3	4	4	4	1	1	1	1	1	1
	2	2	2	3	1	1	1	1	1	2	4	4	4	4	4	4	3	3	3	1	1	1
	3	2	2	3	1	1	1	2	3	3	4	4	4	4	4	4	4	4	4	1	1	1
4	1	2	3	3	2	2	3	2	2	3	4	4	4	4	4	4	4	4	4	2	3	4
	2	3	3	4	2	3	4	3	3	4	4	4	4	4	4	4	4	4	4	2	3	4
	3	4	4	4	2	3	4	3	3	4	4	4	4	4	4	4	4	4	4	2	3	4

Table 2. Risk category of body positions according to relative frequency. Source ("OWAS Method - Ovako Working Analysis System" 2015)

	Relative Frequency	≤10%	≤20%	≤30%	≤40%	≤50%	≤60%	≤70%	≤80%	≤90%	≤100%
BACK	Straight back	1	1	1	1	1	1	1	1	1	1
	Bent back	1	1	1	2	2	2	2	2	3	3
	Back with twist	1	1	2	2	2	3	3	3	3	3
	Bent back with twist	1	2	2	3	3	3	3	4	4	4
ARMS	Two arms down	1	1	1	1	1	1	1	1	1	1
	One arm down and one arm raised	1	1	1	2	2	2	2	2	3	3
	Two arms raised	1	1	2	2	2	2	2	3	3	3
LEGS	Sit	1	1	1	1	1	1	1	1	1	2
	Standing	1	1	1	1	1	1	1	1	2	2
	On a straight leg	1	1	1	2	2	2	2	2	3	3
	On bent knees	1	2	2	3	3	3	3	4	4	4
	On a bent knee	1	2	2	3	3	3	3	4	4	4
	Kneeling	1	1	2	2	2	3	3	3	3	3
	Walking	1	1	1	1	1	1	1	1	2	2

In accordance with the risk identification in each of the tables presented above, Table 3 describes the effect of having for a long time and the corrective action required to avoid harmful problems in the musculoskeletal system.

Table 3. Risk categories. Source ("OWAS Method-Ovako Working Analysis System" 2015)

Risk Category	Risk Category	Required Action
1	Normal and natural posture without harmful effects on the skeletal muscle system.	No action required.
2	Posture with the possibility of causing damage to the musculoskeletal system.	Corrective actions are required in the near future.
3	Posture with harmful effects on the musculoskeletal system.	Corrective action is required as soon as possible.
4	The load caused by this posture has extremely damaging effects on the musculoskeletal system.	Corrective action is required immediately.

The last method to be considered is REBA, which evaluated the postures of the upper and lower limbs of the gas station attendant in a manner that allows the tester to estimate the exposure to risk factors that can cause cumulative traumatic disorders due to dynamic and static postural loading. In order to arrive at the total risk it is necessary for this method to evaluate by groups, group A (Table 4) corresponds to the members of the trunk, neck and legs, and for group B (Table 5) the arm, forearm and wrist. Once

Table 4. Group A REBA.

	Neck											
	1				2				3			
	Legs				Legs				Legs			
Trunk	1	2	3	4	1	2	3	4	1	2	3	4
1	1	2	3	4	1	2	3	4	3	3	5	6
2	2	3	4	5	3	4	5	6	4	5	6	7
3	2	4	5	6	4	5	6	7	5	6	7	8
4	3	5	6	7	5	6	7	8	6	7	8	9
5	4	6	7	8	6	7	8	9	7	8	9	9

Table 5. Group B REBA

	Forearm					
	1			2		
	Wrist			Wrist		
Arm	1	2	3	1	2	3
1	1	2	2	1	2	3
2	1	2	3	2	3	4
3	3	4	5	4	5	5
4	4	5	5	5	6	7
5	6	7	8	7	8	8
6	7	8	8	8	9	9

the combination has been carried out to give the partial risk, the results are combined in Table 6 (NTP 601: Evaluation of working conditions: postural load). REBA (Rapid Entire Body Assessment), (n.d.) method. If necessary, if conditions exist on the type of muscular activity during the task, the risk or final score could be increased by up to 3 points. ("REBA Method - Rapid Entire Body Assessment" 2015.)

Table 6. Group C REBA- Group A and B scores. Source ("REBA Method - Rapid Entire Body Assessment" 2015)

	Score B											
Score A	1	2	3	4	5	6	7	8	9	10	11	12
1	1	1	1	2	3	3	4	5	6	7	7	7
2	1	2	2	3	4	4	5	6	6	7	7	8
3	2	3	3	3	4	5	6	7	7	8	8	8
4	3	4	4	4	5	6	7	8	8	9	9	9
5	4	4	4	5	6	7	8	8	9	9	9	9
6	6	6	6	7	8	8	9	9	10	10	10	10
7	7	7	7	8	9	9	9	10	10	11	11	11
8	8	8	8	9	10	10	10	10	10	11	11	11
9	9	9	9	10	10	10	11	11	11	12	12	12
10	10	10	10	11	11	11	11	12	12	12	12	12
11	11	11	11	11	12	12	12	12	12	12	12	12
12	12	12	12	12	12	12	12	12	12	12	12	12

Table 7 describes the risk and the action to be required according to the final score.

Table 7. REBA Method Risk ("REBA Method - Rapid Entire Body Assessment" 2015)

Score	Level	Risk	Action
1	0	Negligible	No action necessary
2 o 3	1	Low	Action may be necessary.
4 a 7	2	Medium	Action is needed.
8 a 10	3	High	Action is needed as soon as possible.
11 a 15	4	Very High	Action is needed immediately.

Finally, after collecting the results of each method, the respective comparison of the procedures used is carried out so that the corrective actions for musculoskeletal disorders are identified, seeking the improvement of the gas station workstation, leading to a decrease in the overall risk level and/or specific risk agents.

4 Results

In relation to the standardized Nordic questionnaire, applied in order to detect and analyze musculoskeletal disorders, it was found that the motorcycle service assistant worker has had discomfort in the neck and knees for more than a year due to stress and standing posture during his work. However, although he expresses this discomfort in his answers, he admits that it is not an impediment to perform his tasks. For his part, the gas station attendant in charge of the car dispenser said he felt discomfort in his left hand or wrist a week ago due to temperature changes. Additionally, in the last 12 months he has had discomfort that has lasted from 1 to 7 days with episodes of less than one hour. He also described the discomfort as moderately strong, a value of 3; however, it has not been necessary to change position or job since it is not an impediment to perform their work.

The OWAS method was carried out through one hundred observations of tasks carried out by the gas station attendant, which were evidenced by photographic records with a sampling frequency of twenty seconds, that is, the frequency with which the adopted postures will be recorded; thus allowing the measurement in angles of the positions of back, arms and legs with an external load less than 10 kg. In the combination of the different postures during the 100 observations, it was recorded that 86% of these in the gas station attendant of cars scored risk 1 and 14% a risk 2, for the gas station attendant of motorcycles 94% for risk 1 and the rest for risk 2.

Table 8 shows the risk of each position according to the relative frequency, for the gas station attendant of automobiles presents a risk category 1, given that their postures are normal and natural without harmful effects on the musculoskeletal system, as well as for the gas station attendant of motorcycles adding risk category 2 in standing position with the possibility of causing damage to the musculoskeletal system. In addition, the percentage of appearance of each individual posture of back, arms and legs is shown; a greater percentage is found in the posture of two arms down, back straight and standing; which means that the risk of physical load due to rigid postures is greater indicating priority for ergonomic intervention.

Table 8. Result according to the OWAS relative frequency

	Automobile		**RISK**	Motorcycle		**RISK**
Straight back	86	86%	1	94	94%	1
Bent back	13	13%	1	6	6%	1
Back with twist	1	1%	1	0	0%	1
Bent back with twist	0	0%	1	0	0%	1
Two arms down	94	94%	1	84	84%	1
One arm down and one arm raised	6	6%	1	16	16%	1
Two arms raised	0	0%	1	0	0%	1
Sit	0	0%	1	0	0%	1
Standing	60	60%	1	81	81%	2
On a straight leg	17	17%	1	17	17%	1
On bent knees	0	0%	1	0	0%	1
On a bent knee	1	1%	1	0	0%	1
Kneeling	0	0%	1	0	0%	1
Walking	22	22%	1	2	2%	1

In the REBA method, the result given by the combination of the postures in the upper and lower limbs distributed by groups A and B, gives as a final score, including the type of muscular activity performed during the task, the number 5. This score corresponds to a medium risk with a necessary action, focusing on the safety and well-being of the worker. For both workers with attention to different types of vehicles, one for cars and one for motorcycles gives the same level of risk (Figs. 2 and 3).

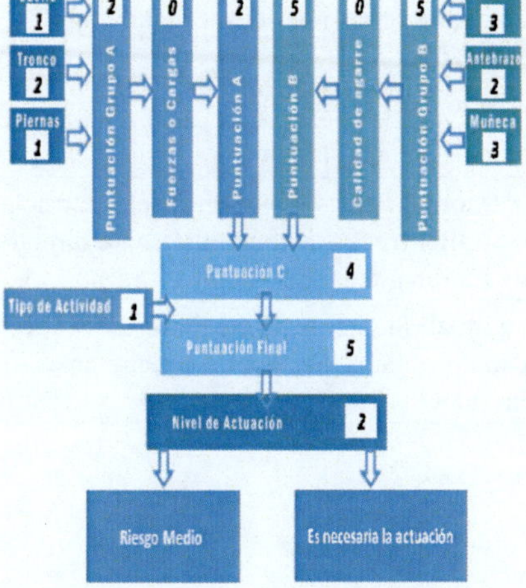

Fig. 2. Risk result motorcycles gas station attendant

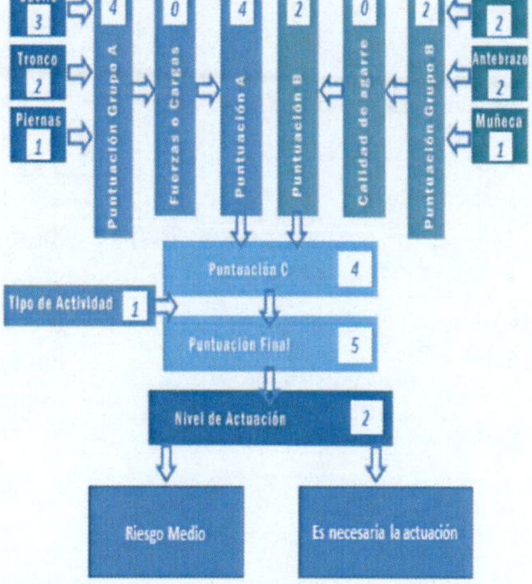

Fig. 3. Risk result automobiles gas station attendant

As for the checklist results, they were analyzed and ranked according to the ten areas covered, referring to the item code, problems and improvement plans evaluated in this method. Table 9 will illustrate the verification of the Check list applied in GS:

Table 9. Results checklist

Área	Códe	Results
Material handling and storage	003	• Tiles in an administrative area produce slippage when wet or humid
	004	• Provide ramps with an inclination of 5 to 8%
	009	• In the oil and lubricants sales office, an elevator is suggested to place products in high areas
Manual tools	025	• Not applicable to the gas station
Safety of service machinery	043	• Not applicable to the gas station
	046	• Not applicable to the gas station
	053	• Not applicable to the gas station
	054	• Not applicable to the gas station
Improved workplace design	062	• The station does not have a resting place for gas station attendants because their work requires permanent standing • After a period of time, the gas station attendants lose the natural posture of their bodies
	063	• The Gas Station does not allow the alteration and change of posture (standing and sitting) to the gas station attendants during their working day
	064	• Chairs or benches are not provided for occasional sitting
Lighting	074	• Improved lighting in storage areas
Workplaces		• It complies with all the items, in other words, it does not contain any problems, nor does it require an improvement plan
Environmental risks	088	• Not applicable to the gas station
	089	• Not applicable to the gas station
	091	• Verify if the external agents vibrations such as: trucks, dump trucks, trailer trucks, among others, are directly affecting gas station attendants health
Hygiene services and rest areas		• It complies with all the items, in other words, it does not contain any problems, nor does it require an improvement plan

(continued)

Table 9. (*continued*)

Área	Códe	Results
Personal protective equipment	098	• No signposting of the use of personal protective equipment in the work area
	101	• The risk that the hydrocarbon fuel may carry by means of the gas station attendant's senses is not taken into account
	103	• The use of safety and protective equipment is not verified
Work organization	112	• No recognition or award for good performance and quality in the work of the gas station attendants
	123	• Hiring is for a specific job and because of this the skills of the person are not taken into account
	124	• Adequacy of bathrooms and ramps for physically handicapped

The checklist identified that GS needs to verify facilities and propose actions for improvement, whether urgent or priority, in the areas of material handling and storage, improved workplace design, lighting, environmental risks, personal protective equipment and work organization.

5 Conclusions

Ergonomics is a fundamental tool in the occupational health of companies, since it allows us to know the risk factors, on which prevention or corrective actions should be considered, through the implementation of ergonomic evaluation methods of workplaces in which musculoskeletal disorders are determined in order to increase the productivity of human talent, reduce physical fatigue, maintain comfort and work stability.

The nature of workplaces in GS is generally linked to the standing posture handled by the gas assistants, therefore, it is necessary to use ergonomic methods that evaluate the postural loads in the upper and lower extremities of the human body such as the Nordic Questionnaire, OWAS and REBA. In addition, perform a global assessment of GS through the Check List.

Thanks to the previously mentioned methods, it is possible to conclude that in the studied GS it is necessary to verify the facilities in the areas of manipulation and storage of materials, making improvements in the illumination and providing ramps with inclination from 5 to 8%. In the office of sales of oil and lubricants, an elevator is also suggested to facilitate the location of products in high areas.

On the other hand, the record of observations for the OWAS method scored in the car gas assistant 86% risk 1 and 14% risk 2 and for the motorcycle gas assistant 94% risk 1 and 6% risk 2; indicating that risk 1 are normal and natural postures without damaging effects on the musculoskeletal system while in risk 2 there is the possibility of causing damage to the system. Likewise, a higher percentage was found in the two-armed down, straight back and standing positions; which means that the risk of physical load, due to rigid postures, is higher indicating priority for ergonomic intervention. Therefore, it is recommended to improve the design of the work stations or to provide rest in a way that workers do not present fatigue or discomfort in their extremities.

Finally, the score generated by the REBA method indicates a medium risk with necessary action in the safety and well-being of the worker, considering taking measures in terms of environmental risks, individual protection equipment and work organization.

References

Anexo 1: Metodología OWAS (2019). http://tesis.pucp.edu.pe/repositorio/bitstream/handle/20.500.12404/6096/CORAL_MARIA_ANALISIS_EVALUACION_CONTROL_RIESGOS_ANEXOS.pdf?sequence=1

El metodo owas para la evaluacion de posturas de trabajo introducción (2019). http://www.fi.uba.ar/archivos/posgrados_apuntes_Metodo_OWAS.pdf

Método LEST - Evaluación global del puesto de trabajo (2015). https://www.ergonautas.upv.es/metodos/lest/lest-ayuda.php. Accessed 19 Aug 2019

Método OWAS - Ovako Working Analysis System (2015). https://www.ergonautas.upv.es/metodos/owas/owas-ayuda.php. Accessed 19 Aug 2019

Método REBA - Rapid Entire Body Assessment (2015). https://www.ergonautas.upv.es/metodos/reba/reba-ayuda.php. Accessed 19 Aug 2019

NTP 601: Evaluación de las condiciones de trabajo: carga postural. Método REBA (Rapid Entire Body Assessment) (2001). https://www.insst.es/InshtWeb/Contenidos/Documentacion/FichasTecnicas/NTP/Ficheros/601a700/ntp_601.pdf

Revista HSEC - OWAS: Un método práctico y simple de evaluación ergonómica (2013). http://www.emb.cl/hsec/articulo.mvc?xid=198&edi=9&xit=owas-un-metodo-practico-y-simple-de-evaluacion-ergonomica. Accessed 19 Aug 2019

Dissemination and Assessment of the Methods and Tools of Strategic Decision Among the Large Companies of the A.M.B.

Elsa Beatriz Gutiérrez Navas(✉) and María Paula Estévez(✉)

Facultad de Ingeniería Industrial, Escuela de Ingenierías,
Universidad Pontificia Bolivariana, Seccional, Bucaramanga, Colombia
elsa.gutierrez@upb.edu.co,
maria.estevez.carvajal.2018@gmail.com

Abstract. The strategic decision processes allow companies to evaluate different alternatives of strategies to implement to achieve a competitive advantage, based on internal and external conditions to the organization. The present study sought to identify the methods and tools to support the strategic decision processes with greater dissemination and assessment among the large companies of the Metropolitan Area of Bucaramanga (Colombia) between 2013 and 2017, for which a bibliographic review was carried out in the Scopus, Science Direct and Google Scholar databases and 58 companies were surveyed, out of a population of 186, managing to exceed the sample for a margin of error of 10% and a confidence level of 90%. The study allowed to identify 35 methods and tools to support the strategic decision processes widely disclosed in the literature, of which 5 adopted by 50% or more of the companies studied are detected, evidencing a lag in the adoption of novel methods and tools. Likewise, the 4 most valued methods and tools were identified for their contribution to the strategic decision processes and the 5 with the greatest dissemination and assessment among the companies studied. The results of the study allow the academy to reflect on the updating of their curricula around the subject and the business sector to identify opportunities to use innovative methods and tools valued for their contribution to strategic decision-making processes.

Keywords: Strategic decision · Strategy · Planning · Methods and tools · Models

1 Introduction

The changing environment in which organizations develop presses a business dynamic characterized by the constant analysis of environmental forces that impose changes in organizations and the development of dynamic capacities and organizational learning processes to anticipate and respond to the forces that alter its organizational balance (Bueno 2016). In this context, as Ackoff reaffirms (quoted by Goodstein et al. 1998), one of the main responsibilities of managers is to decide where organizations should go in the future and how to get there, gaining great importance the strategic management processes.

M. T. Castañeda Galvis et al. (Eds.): ICIE 2019, LNMUINEN, pp. 129–138, 2021.
https://doi.org/10.1007/978-3-030-49370-7_14

Strategic management or strategic administration is the process through which organizations obtain, process and analyze internal information and its context, in order to assess the current situation of the company and be able to anticipate and decide on the path that the organization should follow in the future (Gómez 2006), implementing and evaluating decisions in order to conquer and maintain a competitive advantage (David 2008). In that sense, strategic management encompasses the holistic vision of the organization and its environment, the formulation of measurable and attainable objectives, and specific actions to achieve them, the knowledge of the resources to carry them out, as well as the assertive management of human capital executor of the process (del Canto 2011).

The present study focuses on one of the stages of strategic management called strategic decision, through which it seeks to study different alternatives of strategies and support decision-making against the adoption of one or more of these to position the company in the market, in pursuit of achieving the objectives of survival, growth and usefulness and a competitive advantage in the market (Prieto 2012), making use of methods and tools proposed by different models.

The models correspond to the adaptation of a general theory to a specific restricted context (Zapata-Ros 2015). These constitute a link between the observation of reality and theory, and correspond to representations of a phenomenon or fact, made to interpret it in part or its entirety, through elements such as diagrams, graphs or even an equation (Oliveira 2007). Given that the effectiveness of these models is in their systematic use and their adjustment to the characteristics of the context and of the actors involved, the models establish methodologies to follow, generically referred to as "tools or methods" (Ferradas 2007, p. 39).

In the present study, a bibliographic review of the methods and tools to support the strategic decision processes proposed until 2017 is carried out with greater diffusion in the scientific literature, from which an evaluation of the level of adoption and of valuation of these by the large companies of the Metropolitan Area of Bucaramanga in the years 2013 to 2017. The results of this study sought to highlight the importance of the decision stage in strategic management and allow companies to focus their efforts on the application of methods and tools that have represented major contributions for the companies studied.

2 Methodology

For the development of the work, it was based on a review of the historical evolution of strategic management, using the Scopus and Science Direct databases with the search equations "Evolution Strategy" and "Strategy AND Management" and having as criteria the search for bibliographic reviews published from 2012 to March 2017, not being identified publications of interest for the study. Subsequently, a search was conducted on Academic Google of articles and bibliographic reviews published in the same time range was carried out with the equation "Evolution Strategy", finding 57 documents, of which 7 served for research: Planellas (2016), Planellas and Muni (2015), Castellanos and Cruz (2014), Pulgarin and Rivera (2012), Rivera and Malaver (2011), Labarca (2008) and Saavedra (2005).

In addition to these documents, the books of Introduction to the General Theory of Administration (Chiavenato 1989), Theory of the Organization (Hodge et al. 1998), Strategic Management (Gómez 2014), Strategic decisions: The 30 most useful models (Planellas and Muni 2015), Concepts of strategic administration (David 2008), and Strategic tools: a support for the management decision-making process (Pulgarín and Rivera 2012) to identify the models that up to 2017 have theoretically supported the strategic decision processes.

From the previous review, a timeline was built with 35 methods and tools proposed until 2017 to support the strategic decision, in which methodologies are established to propose or evaluate different strategic alternatives. Based on the above, an analytical questionnaire was constructed, which was applied to those in charge of the strategic planning and administration processes in the large companies of the Bucaramanga Metropolitan Area. Through this, they were asked to identify the methods and tools applied by their companies in the strategic management processes developed between 2013 and 2017 and evaluate their contribution.

To obtain the target population, the large companies with headquarters in the Bucaramanga Metropolitan Area that renewed their commercial registration in 2017 and reported financial information to the Bucaramanga Chamber of Commerce between 2013 and 2017, were taken as the criterion of inclusion, guaranteeing that they had a minimum age of five years. To identify the large companies, those that in December 2017 had assets for more than 30,001 legal minimum monthly salaries were selected, according to Law 905 of August 2 of 2004 of the Republic of Colombia. From this, a population of 186 companies was determined and an ideal sample of 50 companies was calculated for a margin of error of 10% and a confidence level of 90%, and finally, information was collected from 58 of the companies under study.

3 Results

Over the last fifty years, there has been an impressive collective effort to develop the sciences of business administration and management. From the world of organizations and from academia, engineers, economists, sociologists, philosophers, anthropologists and psychologists have taken the time to investigate the complex and exciting phenomenon of the development of organizations and to devise and design models and tools to help managers to face their business challenges and develop strategic management processes (Planellas 2016).

Taking into account that the theories are specified in models for certain contexts and more restricted areas of explanation (Zapata-Ros 2015), and that the models define methodological frameworks that facilitate their practical application, using methods and tools (Ferradas 2007, page 39), the present study was based on a literature review of scientific literature, from which 35 methods and tools proposed until 2017 were identified to support the strategic decision processes in organizations widely disseminated in the scientific literature.

Subsequently, we proceeded to evaluate the level of adoption of these 35 methods and tools by large companies in the Metropolitan Area of Bucaramanga (Colombia), for which a questionnaire was applied to those responsible for the planning and strategic management processes in these. Table 1 shows the number and percentage of companies that have adopted the most widely used methods and tools among the companies under study to support the strategic decision processes, being used by 15% of the companies or more.

Table 1. Level of adoption and contribution of the most widely disclosed strategic methods and decision tools (Own elaboration)

Method/Tool	Companies that apply		Evaluation of the contribution of the method or tool			
	No.	%	1	2	3	Average Rating
Risk analysis	42	72,4%	6	17	19	2,3
Balanced scorecard	34	58,6%	10	16	8	1,9
Management by objectives (MBO)	33	56,9%	10	8	15	2,2
Strategy Maps (balanced scorecard)	31	53,4%	13	16	2	1,6
SWOT matrix	29	50,0%	10	15	4	1,8
Value proposition design	23	39,7%	5	13	5	2
Porter's generic strategies	23	39,7%	7	13	3	1,8
Business model canvas	19	32,8%	4	9	6	2,1
Life cycle analysis	16	27,6%	7	7	2	1,7
Strategic alliances (strategor)	14	24,1%	10	2	2	1,4
Decision trees	13	22,4%	8	3	2	1,5
Strategic prospective	12	20,7%	7	1	4	1,8
Golden circle	11	19,0%	3	5	3	2
Ansoff matrix	11	19,0%	5	5	1	1,6
Three dimensional business definition (abell)	9	15,5%	6	3	0	1,3

Table 2. Origin and usefulness of the methods and tools of strategic decision with greater diffusion (Own elaboration)

Origin	Method or tool	Usefulness
1650	Risk analysis	Risk analysis estimates the probability of undesirable events (risks) and measures their magnitude over a given period, to establish continuous improvement programs and action plans for contingency, emergency or crises (Carson 2008)
1954	Management by objectives (MBO)	A process that seeks to set objectives and goals for each area of responsibility and member of the organization in a top-down manner and then evaluate the results achieved in a bottom-up manner (de Harlez 2016)

(*continued*)

Table 2. (*continued*)

Origin	Method or tool	Usefulness
1982	SWOT matrix	Conceptual scheme to compare external threats and opportunities, together with the internal strengths and weaknesses of the organization, to formulate offensive, reactive, adaptive or defensive strategies (Munuera and Rodríguez 2007)
1992	Balanced scorecard (BSC)	A tool that transforms a company's strategy into a carefully selected set of objectives from four perspectives (Financial, Clients, Internal Processes and Learning and Development) with their respective goals, indicators and action plans that facilitate their management (Kaplan and Norton 2008)
2003	Strategy maps (balanced scorecard)	Graphic representation of the company strategy in which the strategic objectives are set in four perspectives (Financial, Clients, Internal Processes and Learning and Development) with their respective cause-effect relationships, as a basis for the establishment of goals and indicators (Kaplan and Norton 2003)

Table 2 shows the origin and usefulness of the five most widely used methods and tools among the companies under study that support the strategic decision processes, being adopted by more than 50% of these.

As can be seen in Table 2, the origin of the four methods and tools to support the strategic decision-making processes with greater diffusion among the large companies of the A.M.B. dates back 16 years or more, which evidences the delay in the adoption of innovative methods and tools to support strategic decision making in the companies of the region. It is also noteworthy that Risk Analysis and Management by objectives are methods or tools that were proposed in 1650 and 1954, respectively, but are still valid among the companies studied.

Of the 5 methods and tools most widely disseminated among the large A.M.B. companies, three are closely related. On the one hand, the strategic map and the balanced management board were proposed by Kaplan and Norton within the model called Balanced Scorecard, to communicate, operationalize and evaluate the business strategy at all levels of the organization with the support of management measures (Kaplan and Norton 2008). Likewise, this model is based on the conceptual bases proposed by the Management by Objectives, which seeks to deploy the basic purpose and mission of an organization in long- term organizational objectives and subsequently objectives by area, until individual objectives are reached for each member of the organization, in order to achieve their support and commitment to the strategic plans, to subsequently evaluate and measure results in a bottom-up manner, until consolidating the results of the organization (Monllor 1996).

Subsequently, the perception among those responsible for the strategic planning and management processes of the evaluated companies was evaluated on the contribution of the methods and tools adopted in the strategic decision processes to support the evaluation of alternative strategies and decision making in front of which to implement. For this purpose, a scale of 1 to 3 was used, where 1 corresponds to a slight contribution, 2 to a moderate contribution and 3 to a significant contribution. Table 1 shows, for the most widely disseminated methods and tools, the number of companies that rated them on each scale and the average rating obtained. The analysis of the perception of those responsible for the strategic management processes on the contribution of the 35 methods and tools used to support the strategic decision processes, allows us to conclude that those with the highest valuation were: the Transaction Costs Study (3.0), the Eliminate-Reduce-Raise- Create Matrix to formulate a Blue ocean strategy (3.0), the Commercial Review Process or Operational Due diligence (2.8) and

Table 3. Origin and usefulness of the strategic decision methods and tools most valued by those responsible for strategic planning and management (Own elaboration).

Origin	Method or tool	Usefulness
1975	Transaction cost	It seeks to identify the sources of the transaction costs of transferring goods and services along the organizational border and the mechanisms to economize them. It makes it easier to decide which transactions to develop internally, what to buy, what to outsource, what operations to carry out jointly when to integrate vertically or diversify (Salgado 2003)
1993	Operational due diligence	It allows identifying the profit generators of a business through the analysis of four key aspects that determine its attractiveness: market analysis, competitive positioning, client evaluation, and company analysis; to facilitate decision making regarding investment in new businesses and the negotiation of strategic alliances (Martos 2013)
1994	Competing for the future	It raises the fact that organizations seek competitiveness through time, applying three strategies: restructuring the product portfolio and reducing the structure; in-depth revision of processes and their modification through continuous improvement and reinvention of the industry, thus seeking to regenerate their core competencies (Planellas and Muni 2015)
2005	Eliminate-Reduce-Raise-Create (ERRC) Grid	It seeks to question the strategic logic and business model of an industry, identifying which variables to eliminate, reduce, raise or create, in order to build a new value curve that leads the company to a blue ocean strategy, leaving aside the dilemma between differentiation and low cost, and facilitating synthesizing and communicating the strategy (Kim and Mauborgne 2005)

Competing for the future (2.5). However, the level of adoption of these methods and tools among the companies under study did not exceed 7%. The origin and usefulness of these methods and tools are presented in Table 3.

As can be seen in Table 3, of the methods and tools with the greatest value among those responsible for strategic management processes in large companies in the Bucaramanga Metropolitan Area, the first three are focused on orienting decision making in the face of the convenience of entering new businesses, expanding or restructuring operations, outsourcing some activities or developing strategic alliances; while the last two invite strategic innovation processes, reinventing the industry in which one competes or wishes to compete.

Finally, the methods and tools with the greatest dissemination among the companies studied (adopted by 15% of the companies or more) were evaluated, highly valued for their contribution to strategic decision-making processes (See Table 1). The methods or tools that obtained a score greater than or equal to 2.0 on a scale of 1 to 3 were: Risk Analysis (2.3), Management by Objectives (2.2), Business Model Canvas

Table 4. Origin and usefulness of strategic decision-making methods and tools with greater dissemination and valuation (Own elaboration)

Origin	Method or tool	Usefulness
2004	Business model canvas	A tool that makes it easier for organizations to create, develop and capture value by studying nine blocks that answer four questions: What (Value Proposition), Who (Customer Segments, Customer Relationships and Channels), How (Key Resources, Key Activities and Key Associations) and How Much (Income and Costs) (Sierra et al. 2018)
2009	Golden circle model	It establishes three questions that help formulate the mission and strategic core of an organization: Why, How and What. It seeks to make the purpose or reason to be of the company (Why) the one that permeates decisions about its strategies and value proposals (How) and the offer of products and services of the organization (What), inspiring people (Planellas and Muni 2015)
2014	Value proposition design	Method of visual representation consisting of three parts: the customer profile, which details the tasks, frustrations and joys of a customer segment; the value map, which specifies the value proposition with products and services, relievers of frustrations and creators of joys; and finally the fit, which corresponds to the relationship between the above (Osterwalder et al. 2015)

(2.1), Value Proposition Design (2.0) and the Golden Circle Model (2.0). The origin and usefulness of the first two methods and tools are detailed in Table 2 and the last three in Table 4.

This allows us to conclude that the people in charge of strategic planning and management processes value mainly those methods and tools that allow them to make decisions according to the risks they face (Risk Analysis), to think the strategy from the basic purposes of an organization (Golden Circle Model), to stimulate innovation processes in their business models and value proposals to attend certain market segments (Business Model Canvas and Value Proposition Design) and to deploy the objectives in all levels of the organization (APO).

4 Discussion

The study allowed the identification of 35 methods and tools highly divulged in scientific literature and strategic management books proposed until 2017 to support strategic decision-making processes. Of these, the five methods and tools most widely adopted by large companies in the Bucaramanga Metropolitan Area between 2013 and 2017, when applied by 50% or more of the companies studied, were: Risk Analysis, Management by Objectives, SWOT Matrix, Balanced Management Board and Strategic Map; the latter two proposed by Kaplan and Norton through the Balanced Scorecard.

The methods and tools adopted by more than 50% of the large companies in the Metropolitan Area of Bucaramanga (Colombia) originated 16 years or more ago, which shows a lag in the teaching of innovative methods and tools by academia and in the adoption of these among the companies studied. These are focused on identifying the main risks faced by the organization to establish contingency, emergency or crisis plans (Risk Analysis), devise and evaluate strategic alternatives according to the conditions of external and internal environment of the organization (SWOT), as well as deploy the strategy at different levels of the organization and evaluate its achievement through objectives, indicators and goals (APO, Strategic Map and Balanced Management Board).

Despite their low diffusion, when implemented by less than 7% of the companies under study, the methods and tools most valued by those in charge of strategic planning and management processes for their contribution to the strategic decision phase were: the Transaction Costs Study, the Eliminate-Reduce-Raise-Create Matrix, the Commercial Review Process and Competing for the Future; which focus on orienting decisions at the corporate level to enter, leave or stay in business and stimulate innovation processes, with a view to rethinking and reinventing an industry.

Among the methods and tools most widely disseminated in the companies evaluated, those best valued for their contribution to strategic decision-making processes were: Risk Analysis, Management by Objectives, Business Model Canvas, Value Proposition Canvas, and Golden Circle Model. These allow companies to identify and assess the risks they face, make strategic decisions aimed at mitigating those risks and innovate their business model and value proposition without losing sight of its basic purpose, to finally deploy the strategy at all organizational levels.

References

Bueno, E.: Dinámica empresarial y creación de ventajas adaptativas en el desarrollo sostenible. Revista de la Asociación Española de Contabilidad y Administración de Empresas AECA (115), 3–5 (2016). http://www.aeca1.org/revistaeca/revista115/115.pdf

Carson, B.: Corre el riesgo: Aprenda a identificar, elegir y vivir con un riesgo moderado. Vida, Michigan (2008)

Castellanos, J., Cruz, M.: Una mirada a la evolución histórica de la estrategia organizacional. Estudios Avanzados de Liderazgo **1**(3), 28–51 (2014)

Chiavenato, I.: Introducción a la Teoria General de la Administración. McGrawHill, México (1989)

David, F.R.: Conceptos de administración estratégica. Pearson Educación, Naucalpan de Juárez (2008)

de Harlez, R.: Administración por objetivos: Alcance sus metas con empleados motivados. 50Minutos.es (2016)

del Canto, E.: Gerencia Estratégica y Capital Humano. Su prospectiva en los gobiernos. Revista Ciencias Estratégicas **19**(26), 171–184 (2011). https://www.redalyc.org/pdf/1513/151322415003.pdf

Ferradas, P.: Metodologías y herramientas para la capacitación en gestión de riesgo de desastres. Soluciones Prácticas, s.l. (2007)

Gómez, H.: Gerencia estratégica. Panamericana Editorial Ltda, Bogotá D.C. (2006)

Gómez, H.: Gerencia estratégica. Panamericana Editorial Ltda, Bogotá D.C. (2014)

Goodstein, L., Nolan, T., Pfeiffer, W.: Planeación estratégica aplicada. Como Desarrollar un plan que realmente funcione. MacGraw-Hill Interamericana, S.A., Bogotá (1998)

Hodge, B., Anthony, W., Gales, L.: Teoría de la organización: Un enfoque estratégico. Pearson Prentice Hall, Madrid (España) (1998)

Kaplan, R.S., Norton, D.P.: Strategy Maps: Converting Intangible Assets into Tangible Outcomes. Harvard Business Press, Boston (2003)

Kaplan, R., Norton, D.: The Execution Premium. Grupo Planeta, España (2008)

Kim, W., Mauborgne, R.: La estrategia del océano azul. Norma, Bogotá (2005)

Labarca, N.: Evolución del pensamiento estratégico en la formulación de la estrategia empresarial. Opción **24**(55), 47–68 (2008)

Martos, A.M.: La due diligence financiera. Profit, Barcelona (2013)

Monllor, J.: Economía, legislación y administración de empresas. Editum, Murcia (1996)

Munuera, J.L., Rodríguez, A.I.: Estrategias de marketing: Un enfoque basado en el proceso de dirección. ESIC, Madrid (2007)

Oliveira, G.E.: Modelos teóricos aplicados al turismo. Estudios y perspectivas en turismo (2007)

Osterwalder, A., Pigneur, Y., Bernarda, G., Smith, A.: Diseñando la propuesta de valor: Cómo crear los productos y servicios que tus clientes están esperando. Grupo Planeta, Barcelona (2015)

Planellas, M.: La evolución de la estrategia. Harvard Deusto Bus. Rev. (257), 6–16 (2016)

Planellas, M., Muni, A.: Las decisiones estratégicas: Los 30 modelos más útiles. Penguin Random House Grupo Editorial, Barcelona (2015)

Prieto, J.: Gestión estratégica organizacional. Ecoes Ediciones, s.l. (2012)

Pulgarín, S.A., Rivera, H.A.: Las herramientas estratégicas: un apoyo al proceso de toma de decisiones gerenciales. Universidad Libre, Bogota (2012)
Rivera, H., Malaver, M.: ¿Qué estudia la estrategia? Universidad del Rosario, Bogotá D.C. (2011)
Saavedra, J.: Administración estratégica: evolución y tendencias. Economía y Administración (64), 61–75 (2005)
Salgado, E.: Teoría de costos de transacción: una breve reseña. Red de Revistas Científicas de América Latina y el Caribe, España y Portugal, pp. 61–78 (2003)
Sierra, H.A., León, L.J., Muñoz, M.A., López, P.: Propuesta de emprendimiento social FUNEPU. Universidad Nacional Abierta y a distancia-UNAD, Bucaramanga (2018)
Zapata-Ros, M.: Teorías y modelos sobre el aprendizaje en entornos conectados y ubicuos. Bases para un nuevo modelo teórico a partir de una visión crítica del "conectivismo". Sistema de Información Científica, pp. 69–102 (2015)

Logistic - Supply Chain - Sustainability

Application of the BPM Life Cycle in the Post-harvest Process in a Company in the Floricultural Sector

Claudia Rodríguez(✉), Diego A. Dávila, and Nicolás Quintero

Faculty of Industrial Engineering, Escuela Colombiana de Ingeniería Julio Garavito, Bogotá, Colombia
claudia.rodriguez@escuelaing.edu.co,
diego.davila-v@mail.escuelaing.edu.co,
nicolas.quintero-p@mail.escuelaing.edu.co

Abstract. In the constant growth of the floricultural industry in Colombia, the continuous improvement of processes requires the implementation of new models and technologies that allow to keep this market efficient and sustainable Therefore, Business Process Management (BPM) has proven to be appropriate for increasing the performance and competitiveness in the main processes of an organization. The life cycle was applied to the post-harvest process, considering that the scope of this study reaches the proposal of redesign and improvement proposals, showing that this study is valuable, since in general the flower growing companies have the same problems and disadvantages. Finally, this article presents improvements that can be used by companies in this sector.

Keywords: Business Process Management (BPM) · BPMN 2.0. notation · Lifecycle · Postharvest of flowers

1 Introduction

Process documentation in any organization is vital to find the problems that exist in each area of the company, so in this article we will talk about the life cycle for the structure that manages BPM.

First, there will be a review of the literature that covers flower exports in Colombia, documentation and notation in BPMN 2.0, and the life cycles to find the problems within the company.

Therefore, the work was carried out in a company in the floricultural sector to determine the status of its processes in accordance with the system offered by BPM, leading to certain results with the application of the life cycle, to finally conclude with the state the company is in.

M. T. Castañeda Galvis et al. (Eds.): ICIE 2019, LNMUINEN, pp. 141–148, 2021.
https://doi.org/10.1007/978-3-030-49370-7_15

2 Methodology

A systematic search for information was made in scientific articles, which guided the research with a rigorous contextual description of the topics that should be addressed. The articles were selected to identify, differentiate and choose the material, based on the theoretical criteria of the research and thus determine the key concepts that allow establishing which aspects favor the GMP culture in a work team (Argilaga 1986, Gómez and Pimiento 2013).

In addition, scientific databases were consulted, such as ScienceDirect, JSTOR, Emerald, ProQuest, and EBSCO. The keywords used were: BPM, teamwork, culture, communication, leadership, empathy, performance, mental models.

3 Theoretical Framework

The beginning of the floricultural export began in the 60s with the creation of the company "Flores Colombianas Ltda.", which was created by Mr. Edgar Wells who undertook a trip to the United States to study the technical procedures of the crop. There, he scouted the potential of this sector and decided to return to Colombia to create such a company, and on October 18, 1965 they sent the first shipment to the United States reaching a value of US $ 20,000 (Ángel and Rangel 2000, Yaniver and Gomez 2018).

In Colombia, flower production dates back to the thirties with a group of farmers who found a business opportunity in flowers. They led the creation of greenhouses and nurseries in gardens that allowed the land to be plowed to sow flowers, paving the way for the flower market to be one of the main economic activities. Today, the country is positioned as the second largest exporter worldwide after the Netherlands, acquiring competitive advantages due to quality standards that ensure the best sizes, colors, and conservation of a great variety of flowers (Cárdenas and Rodríguez 2011, Rodríguez 2015, Yaniver and Gomez 2018).

Post-harvest is one of the main processes to consider before exporting. Since this depends on the duration, quality, and standards that the customer handles and demands, it requires the constant updating of the concepts of flower treatment after its arrival from cultivation, such as genetic, agronomic, and environmental factors (Pardo 2010).

The processes that comprise the areas of the floricultural sector show great opportunities to improve. for continuous growth. BPM emerges as an improvement alternative which is defined as an integrated process-based management system, which uses specialized information systems and covers all operational and business processes, ensuring that the organization is more productive and competitive, fostering continuous improvement in line with the company strategy (Rodríguez 2015).

To understand the implementation of BPM in a company, it is necessary to associate two fundamental concepts that allow generating and applying improvements in processes, such as management and technology. The former is defined as the way through which objectives are established and they project resources, and the latter as a system that manages to control and operate the interrelation of the business processes found in the company. To consolidate these two concepts within the company, organizational culture plays a key role, since it creates an obstacle identified as *resistance to change*. Thus, improvement and coupling courses of action must be created for the effective implementation of said tool (Rodríguez 2015, Huergo 2018).

It is necessary to identify how BPM is going to be implemented, which generates a structured vision of how a process can be managed through a concept called life cycle. As stated, it is carried out through a cycle that seeks continuous improvement and the implementation of each concept while maintaining the line of process interrelation, given to any organization (Dumas et al. 2013).

Therefore, the steps to follow for the application of the life cycle are covered as follows (Fig. 1):

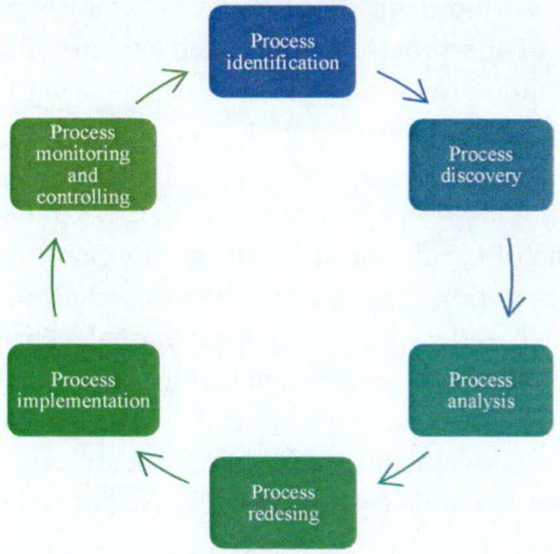

Fig. 1. Life cycle phases. Source: Dumas et al. 2013.

One of the main bases for the identification of certain life cycle factors is the interrelation of processes within the organization or specifically in an area of it. This allows to create documentation where they can analyze and create courses of action for improved continuation of each of the processes; therefore, BPMN 2.0 notation is entered.

This notation has recently emerged as the industry standard notation for modeling business processes, providing a notation that is easily understandable to all business users. It ensures that the processes covered by the BPM initiative lead to consistently positive results, generating maximum value to the company in its business processes (Dumas et al. 2013, Jankovic et al. 2015).

4 Results

Life cycle phases of the post-harvest process

1. Process identification.

The main postharvest processes were identified, from the reception of the flower that comes from the cultivation area until its deployment for export according to the order, where the product is selected and placed at disposal. Next, the main processes that are performed on the flower are identified:

- ✓ Reception.
- ✓ Initial hydration.
- ✓ Classification.
- ✓ Assembly.
- ✓ Final hydration.
- ✓ Packaging.

2. Process Discovery.

This part includes the detail of each of the processes, the way in which they are performed, and the functions that each of the employees performs. A detailed understanding of their workings was performed in each area and through interviews and observation of the process, the main problems were identified.

The documentation of the process was carried out in BPMN 2.0 notation, which shows the sequencing and interrelation of processes, showing the structuring of the life cycle for BPM implementation.

3. Process analysis.

The structuring of the life cycle for the implementation of BPM after identifying the processes and their interrelation through their documentation, continues with process analysis techniques which reflect how each operational activity is behaving to later evaluate them. Its objective is to create an action plan that develops a continuous improvement of the whole process.

In the case of a company in the floricultural sector, we will use the qualitative techniques analyzed from the added value, root cause and problem registration; Below are the techniques:

- Value-added analysis by operation: each of the operations were classified by value-added (VA), business value-added (BVA) and finally those that do not add value (VAN), (Dumas et al. 2013) (Table 1):

Table 1. Value added analysis

Operation	In charge	Classification
Receive flower box	Reception leader	BVA
Register by reader	Reception operator	BVA
Tag boxes	Reception operator	BVA
Receive boxes	Hydration operator	VA
Put in hydration tubs	Hydration operator	VA
Spray flowers	Hydration operator	VA
Place time tag	Hydration operator	BVA
Store in cold room	Hydration operator	VAN
Receive boxes	Sorter	BVA
Sort flowers	Sorter	VA
Send to national flower	Sorter	VAN
Place in lyre	Sorter	VA
Bouquet assemblance	Assembler	VA
Place on conveyor belt	Assembler	VAN

(*continued*)

Table 1. (*continued*)

Operation	In charge	Classification
Receive bouquet	Final band operator	VA
Cut stems	Final band operator	VA
Put in tubs	Operator final band hydration	VAN
Place time tag	Operator final band hydration	BVA
Move to cold room	Operator final band hydration	VAN
Receive tubs	Packer	BVA
Organize tubs	Packer	BVA
Take to storage area	Packer	VAN
Keep in cold room	Packer	VAN
Fill in formats	Shipping packing leader	BVA
Upload order to truck	Packer	VAN

Source: Own elaboration based on life cycle analysis

Percentage of each classification:

- BVA: 36%
- VA: 32%
- VAN: 32%

- Root cause analysis: from a fishbone as a root cause diagram, an analysis was carried out as a result of the problems found in the floricultural sector company analyzing the causes that generated the main reprocessing problem.

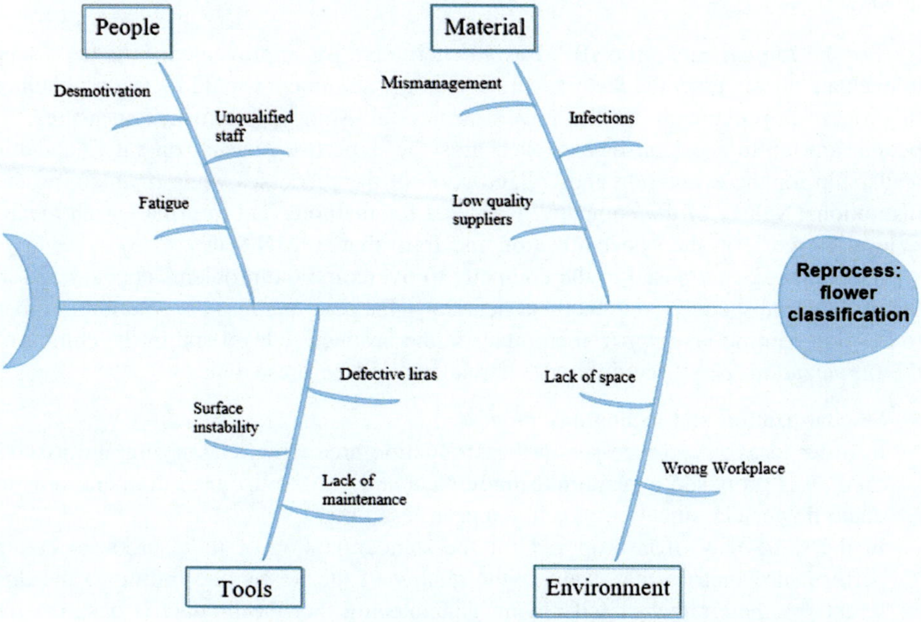

Source: Own elaboration based on life cycle analysis

- Problem registration: the main problems that the company has in the post-harvest process are shown below (Table 2).

Table 2. Problem registration

Name	Explanation
Reprocesses in production area	In the post-harvest process in the production area, there are failures in the classification of the flower, since the moment the flower is assembled there are diseases that generate a return to the classification area, creating constant reprocesses that create unproductivity
Loss of flower to national	The quality standards that the company manages are high, so there are several flowers that get redistributed to the national flower market, which creates unproductive times, as flowers are processed but not exported
Cold room stock	A high amount of inventories is presented in high seasons due to the demand for flowers in the United States; which leads to the accumulation of the inventory in the post-harvest process due to a lack of space or exits of flowers to export
Idle times	The unproductive times are mostly reflected when an order is placed and the flowers are not found to be able to assemble the bouquets and make them available for cold room and packaging

Source: Own elaboration based on life cycle analysis

4. Process redesign

For the implementation of BPM as an alternative for improvement, it is necessary to evaluate the six essential factors, where strategic alignment should focus on planning in process improvement, in the government establishing performance measures, in people and culture established in each area the expertise of each operator, and his leadership for the evaluation and collaboration of the processes while maintaining the institutional values of the company. It is vital for methods and information and technology, to work on the systematization and form that BPMN states.

The approach proposed to the company to overcome the problems encountered in the floricultural sector is to implement heuristics that lead to a feasible solution in order to develop continuous improvement plans at the productive level and in the culture of the organization. Next, the different alternatives will be presented:

- Standardization and technology process
  In order to avoid reprocesses in the production area when classifying and assembling, it is proposed to acquire technological and easy-to-use measurement tools to make flower classification effective and in less time.
  In the same way, it is proposed for the standardization of these tasks, to create performance metrics that evaluate the quality of the flowers according to the different parameters that classify them, guaranteeing that the product is destined for export and not for national flower.

- Parallelism
  Planning the flower supply system by the organization to the crop company is essential to avoid idle times, therefore, in the area of hydration and classification there must be an indicator or alert that reports that the stock in this process is running out to carry out its provisioning and work in parallel between production and supply.
- 5s' methodology
  This methodology seeks to optimize the state of the work environment, facilitate the work of employees and enhance their ability to detect problems, improving productivity, quality in the process and reducing costs and expenses in different areas. The 5 s' are divided into classification, organization, cleaning, standardization and discipline, and habit (Flores 2017).

5 Conclusion

The life cycle of the processes posed by structuring the application of BPM is efficient, since it allows identifying, analyzing and restructuring any process. It must be noted that during the development of this process, improvement plans are being devised from the operating sector of the organization.

The documentation allows a clear understanding of the processes where there was a high interrelation of processes throughout the postharvest, finding alternatives for improvement over the main problems that the company shows. To achieve such goal, a redesign of the process is proposed at the moment of receiving the flower to avoid unproductive times when the process passes through classification, increasing productivity in the number of stems per hour that operators perform.

The application of qualitative life cycle techniques encompasses main factors within the process thanks to the identification of causes of the problems that the company is going through; this enables shock plans to be constantly created to impact each problem.

References

Ángel, M., Rangel, J.: Comercialización internacional de flores, Universidad nacional de Colombia, p. 278 (2000)

Argilaga, T.A.: La Investigacion Cualitativa. Educar **10**, 23–50 (1986)

Cárdenas, M., Rodríguez, M.: Colombia y la creación de una empresa productora, p. 166 (2011)

Dumas, M., et al.: Fundamentals of Business Process Management, 1st edn. Springer, Heidelberg (2013)

Flores, W.: Análisis y propuesta de mejora de procesos aplicando mejora continua, técnica SMED, y 5S, en una empresa de confecciones. Pontificia Universidad Católica del Perú (2017)

Gómez, L.S.S., Pimiento, N.R.O.: Una revisión de los modelos de mejoramiento de procesos con enfoque en el rediseño. Estudios Gerenciales **28**(125), 13–22 (2013). https://www.icesi.edu.co/revistas/index.php/estudios_gerenciales/article/view/1524/html

Huergo, J.: Los Procesos de Gestión (2018). http://servicios.abc.gov.ar/lainstitucion/univpedagogica/especializaciones/seminario/materialesparadescargar/seminario4/huergo3.pdf
Jankovic, M., et al.: Enhancing BPMN 2.0 informational perspective to support interoperability for cross-organizational business processes. Comput. Sci. Inf. Syst. **12**(3), 1101–1120 (2015)
Pardo, F.A.C.: Estado del arte de la poscosecha de flores en colombia, p. 38 (2010)
Rodríguez, C.Y.: Qué es business process management BPM. Revista de la Escuela Colombiana de Ingeniería **25**(98), 23–29 (2015)
Yaniver, E., Gomez, L.: Análisis de la exportación de flores colombianas durante el período de 2012–2017. Universidad Cooperativa de Colombia (2018)

Design of a Storage System for Dynamic Load Management Proportional to the Logistic Operation of a Port Warehouse, Based on the PERT Tool, Optimization and Reliability Models

Harold Enrique Cohen Padilla(✉),
Martha Sofía Carrillo Landazábal(✉), and German Herrera Vidal(✉)

Faculty of Engineering, Industrial Engineering Program, Fundación Universitaria Tecnológico Comfenalco, Cartagena, Colombia
{hcohen,herreravg}@tecnocomfenalco.edu.co,
invest.industrial@tecnologicocomfenalco.edu.co

Abstract. This research work evaluates the management of port logistics, the storage process in miscellaneous warehouses, focusing on network operational analysis (PERT) and its reliability (Reliability), determining from a statistical point of view a model of linear fit by least squares between the waste derived from the real process and a functionally simulated model (black box model) Aigües et al. (2014). The project is based on a descriptive quantitative type of methodology where the probability of the internal process in the winery and the planning/execution of the process is shown, versus variables such as expertise, routine and even the influence of stress due to fiscal and/or mental fatigue as a consequence of port logistics tasks. At the end of the work, a working model based on the handling of these variables is proposed in order to improve the availability of the operative resource and even the response rate from the levels of execution of the tasks.

Keywords: Port logistics storage · Dynamic load management · Reliability · PERT · Optimization models

1 Introduction

A port operator, according to the author Rahma (2019) is the company that provides services in the ports, directly related to the port entity. It is so important that the companies providing these services are in a process of improving their processes and at the forefront in search of favorable results in a highly competitive market.

In recent months there has been an increase in the number of national and global port activities that are present in the ports of the north coast of Colombia, which added to the good trajectory and excellent results of the port operator of this study, have resulted in an increase in contracted services, not without first realizing that this has hindered in some cases or delayed in others, the proper planning of tasks causing a significant increase in delays or non-conformities in the process without being

M. T. Castañeda Galvis et al. (Eds.): ICIE 2019, LNMUINEN, pp. 149–158, 2021.
https://doi.org/10.1007/978-3-030-49370-7_16

transcendental in the provision of service, if they cause discomfort in the relationship between the Port and its users, if they cause discomfort in the relationship between the Port and its users (Fig. 1). That is why we intend to make the analysis of the situations of the context so that measures are taken to help identify relevant and effective solutions in each case, limiting the occurrence of reprocessing, loss of time due to operational failures which increase the response time of the tasks, costs and sometimes possible loss of customers due to deficiencies in the provision of service, delays or non-compliance with agreed procedures and ultimately avoiding the emergence or growth of situations such as low rating of the service, as documented by Pastor (2017) in "Notes on the port handling contract".

Fig. 1. Operations delivered out of time between 2018 and 2019

2 Methodology

In order to carry out a process that would provide a solution to the problems addressed, a case study diagnosis is established (port operator company), taking as reference the standard working conditions, a probabilistic analysis of times is carried out (optimistic, pessimistic and standard from the PERT methodology: Program Evaluation and Review Technique), Vergara and Teixeira (2017) from which a hypothesis is established that, if accepted, the black box model is applied, which is none other than the comparison of the waste obtained from the difference between the working results of the real model and the statistically estimated data of a model simulated in order to be able to make adjustments that lead to a decision-making process that favours the execution of the logistical tasks of loading or unloading containers in the warehouses of the ports, Montero and Sarmiento (2017).

The PERT (Program Evaluation and Review Technique) tool is a continuous, step-by-step process in which the sequences of activities of a planned process are operationalized in which time is the element that allows establishing the precedence of one task over another. This planning and network of tasks can be visualized through a diagram called "network", Santoyo and González (2017) that facilitates the estimation of critical tasks and with it the control of the same throughout the duration of the project.

In the case of the present study, a framework is established that starts from the context of the problematic situation to the validation of the mathematical model, which will allow to have a linear approximation in the recommended and optimal way of operationalizing the tasks to be carried out in the workplace (Fig. 2):

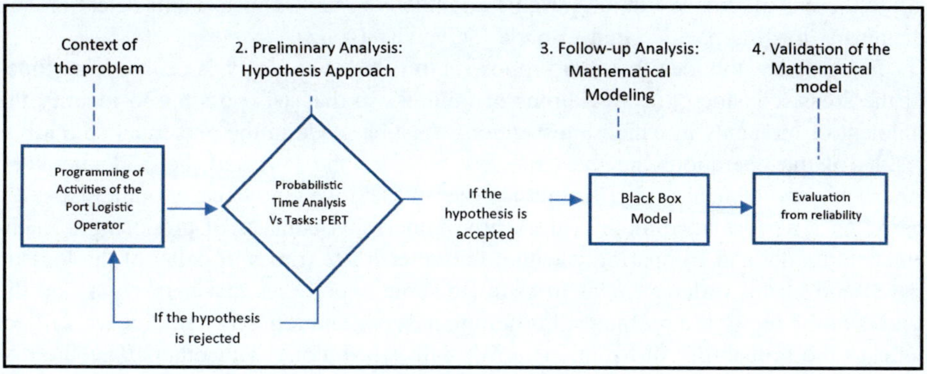

Fig. 2. Model proposed as methodology

To start the process, the first thing to do is to gather basic information about the project and its activities in order to generate a timeline and reference budget. The information to be collected includes the list of activities, the order of precedence of the activities, the durations of activity and the costs of activity. Following the standard PERT methodology, activity durations are provoked as a triplet of optimistic, more probable and pessimistic durations. These form beta distributions to sample later in the analysis. Similarly, activity cost estimates can be taken as optimistic, more likely, and pessimistic costs sampled as triangular distributions. Based on the information gathered, a baseline project network can be built. PERT assumes that the project can be broken down into a set of individual activities represented by a network formed by nodes and arcs, mathematically represented as a cyclical graph directed in which they represent the set of project activities with a certain duration, and the relationships of precedence between them. Depending on the structure of the graph and the activity durations modeled as beta distributions, a critical path can be found in such a way that "a delay of an activity on the critical path will cause a delay in the project as a whole", Collier and Hendrickson (2018).

After analyzing the data and establishing results according to the PERT tool for analysis of the conditions of the tasks and the susceptibility to change or determination of criticality of them in order to contribute to the supervision and control of operations, it is intended to establish a criterion of functionality or reliability of the process, so that through the reliability, as said Mohamed (2013), we can determine the state of the port operations of the company under study so that they can model their trend of behavior to proceed in the planning and subsequent execution of tasks, so that these provide greater efficiency and responds to the current needs of the contractor. In this order of ideas, the situation would be presented according to the black box model, complemented with a statistical analysis, based on the times of operations and the delays reported in an accumulated manner. Black box models are "those that calculate the output variables from a series of mathematical operations with the input variables, without these operations being based on the physical structure or type of process they model. Using these system identification techniques, it is possible to find a list of input variables that allow the calculation of output variables. Predictive maintenance. Fault detection and diagnosis. Las Palmas de Gran Canaria University (2007).

In summary, the idea that you propose is to propose an analysis of the delay times of the areas (considered as deviations or failures) so that it is possible to identify the moment or moments in which interventions must be made in the personnel so that the results of the operations are the expected ones, for the Port and the port operators, Rodriguez and González (2017). In this order of ideas we plan to carry out a first study based on a performance model (black box model) that consists of proposing a mathematical model and comparing it with a failure estimate (times of delay of the logistic areas) and in this order of ideas to estimate some hypotheses that help us to find the causes of the registered problems. For failure analysis and reliability studies we will not refer to the probability distribution of Weibull, since along with others "are the distributions most used to analyze life cycle data, as said Nassar et al. (2018). At the end of the exercise, if possible, a model of optimization of allocation resources will be established by means of linear programming in such a way that the operating conditions of the company are potentialized without having to punish monetarily the way in which tasks are currently executed in the port, Gonzalo San José (2016) (Fig. 3).

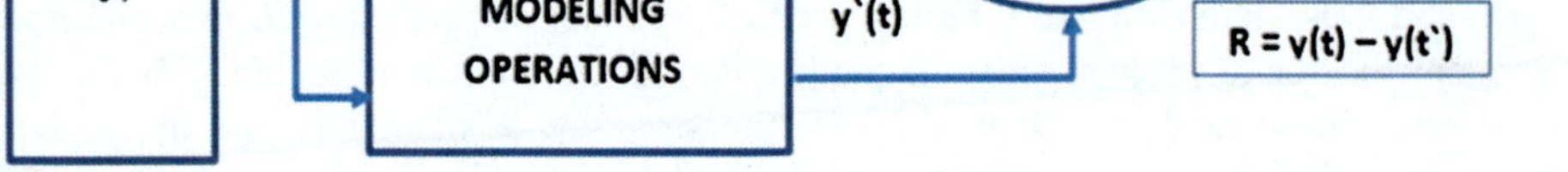

Fig. 3. Black box model

3 Results

The operation in port, as far as it could be seen, depends on factors associated with several work fronts: different companies, different people and different responsibilities, that even assiduously work in coordination to obtain the best results. That is why it becomes a little complicated to optimize the operations of a single front of work expecting global results, but beyond that work is done with the firm intention of improving. In order to establish a working route that limits the variations or looseness in the operations, the critical path is determined or in other words the tasks are found that cannot have delays in the process because they would affect the final result. This action is carried out with the support of the GAMS software. The General Algebraic Modeling System (GAMS) is a tool that is specifically designed to model both linear, nonlinear and mixed optimization problems, and to perform mathematical programming. In the case studied, 24 programming activities are detailed, ranging from the organization of the work order, to the delivery of the documentation that supports the goods on the shelves of the warehouse (when emptying containers) or in the cargo vehicles or dispatch yards if the task performed is a filling of containers.

The analysis of the tasks according to the PERT tool, probably speaking and according to the result in GAMS (Fig. 4), indicates that the critical path corresponds to the activities: A-G-H-L-M-O-P-Q-R-S-U-V-W-X, i.e. the tasks: to receive the documentation of disembarkation of the container, to be located physically and adequately in planning the tasks, to mobilize the container, to locate the container in the site, to revise the seals of the container, to document evidences and to open the container, to analyze the load, to tie and to hoist the loads, as well as to transport the load to the site, to be located in the suitable shelf, to deliver the papers in the office of the auxiliary of the port, final inspection and to close the documentation of delivery with an approximate time of 1983 min by container, which gives us a point of reference for the schedules of the next movements; Therefore to be clear that these are critical activities implies to maintain a stricter control over them so that no deviation can affect them and consequently provide an inadequate service.

For being a probabilistic process, an analysis of conditions of the process is made so that different scenarios could be simulated in the present study, or the results thrown by the software was that the process is not susceptible of adjustments due to that the probabilities of delaying less time of the planned one is practically zero; this due to the rigidity of the procedure in the port and the quantities of companies participants in the process, in which all by regulation and strict control of merchandise does not allow changes in the way to develop the tasks. Due to this situation, a regression model is proposed (linear in this case) to try to deduce the behavior of the process and through the tool proposed and already commented (black box) determine if the residues or differences between the programmed in the model and the data of the operation allow to establish a route or route to follow. Likewise, this condition of internal polarization shows that the optimization of the process can only occur if we complement the work strategy with the tasks of external logistics of containers and handling of operators in the area of exchange of godos.

To continue with the analysis of results, we proceed according to the black box model to establish a mathematical model of regression. Therefore, when doing the practical exercise with reference to a working day in the given cargo terminal, we find that the lost times in each of the work slots are accumulated to determine the "loss time" or lost times of the day and thus determine, with a good analysis of causes, the factors that allow us to identify the factors that trigger a bad service (see Fig. 5).

```
SET  Activity / A*X/;
ALIAS  (Activity,i,j);
SET  Prec (i,j) /
   A, (B,G) , B.(C,D,K),D.E, E,F, G.H, (D,C).I,
   I .J, (D,H) .L, L,M (F,K) .N, M.O (O,N,J) .P, P.Q
   Q.R, R. (S,T), (T,S) .U, U.V, V.M, W.X/;
PARAMETER Duration (Activity)
/A 189.0, B 59.5, C 67.5, D 126.7, E 123.7, F 172.0,
G 618.8, H 181.2, I 118.5, J 134.8, K 120.3, L 58.8,
M 78.8, N 121.7 O 58.0, Q 120.3, R 148.0,
S 117.8, T 60.0, U 67.3, V 126.8, X 161.3/;

FREE VARIABLE  Time;
NONNEGATIVE VARIABLE  S(i);
EQUATIONS  Ctime (i)
                Ptime (i,j)  ;
Ctime ( i ) . .  Time  =G=  S(i) + Duration(i) ;
Ptime(Prec ( i,j ) ) . . s( i ) + Duration( i )  =L= S( j );

MODEL Schedule  /ALL/;
SOLVE Schedule using lp minimizing Time;
DISPLAY Time.1, S.1;
```

Fig. 4. Critical Path of probabilistic process activities. Programmed in GAMS

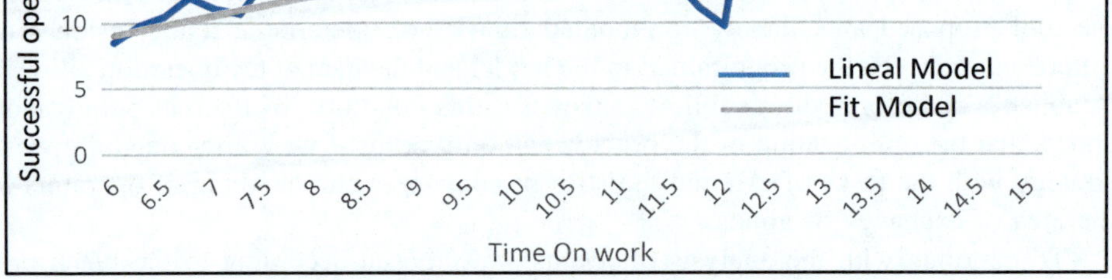

Fig. 5. Model of regresion: actual and linear fit

Residue Analysis
Slope : 1,5034476
Intercept : 0.18864
Arithmetic mean: 5,789760417
Standard deviation :2,525649748
Parameter a (99% confidence level): 2.576
R^2: 0.7370225
R :0,85858

Given the simple regression data Y = mX + b, for all independent "X" variables and all "Y" as a dependent variable, it could be determined that the correlation coefficient, R, indicates that there is a strong or important dependency ratio, in this linear case, of the data studied and the model, since its value is between 0.7 and 0.9. The determination coefficient, R2, indicates that 73.7% of the data are linearly correlated, which leads us to believe that the responses found indicate that the sample behaves according to the Weibull density function (Fig. 6).

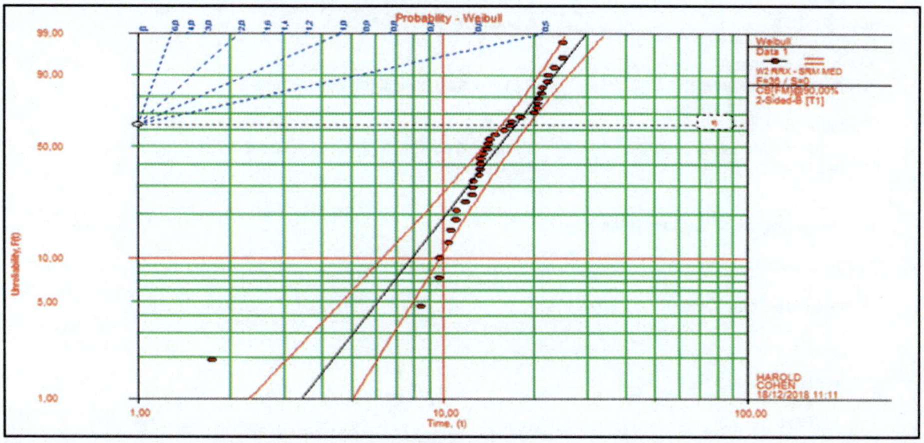

Fig. 6. Probability of the operational time reliability function. Model Programmed in Weibull+.

The density of the process, analyzed with the probability distribution Weibull could be established from the model, that the trend of the behavior of the accumulated times of failures in the provision of the service, has a parameter beta (β) or corresponding to a value of 2.83, which would place it exponentially in the zone of wear or useful life in potential faults due to exhaustion, a point that corroborates that these are moments in which operators are vulnerable to faults due to working hours, which is why it is recommended to take measures, active breaks, rotation of personnel or greater supervision to minimize the consequences of this situation. See more in NPT 331: Fiabilidad: La distribución de Weibull.

In order to finish the process and taking into account the internal rigidity of the process (previously explained by probabilistic PERT model) it is decided to establish a model of assignment of tasks in the logistic exchange relationship in the doors of the warehouses: internal external so that if by optimizing the external work and establishing a work mechanization the failures, the response times and with it the satisfaction of the client can be reduced. This type of problem is called Schedule Optimization Problem and its approximate or heuristic solution is a good answer with respect to the efficiency criteria without the need to be the best. It is a problem in which there will be m machines in parallel where the work to be done must meet some relationships of precedence and the goal is to minimize the completion time of all tasks. Taha (2004). In that order of ideas an objective function and restrictions are established to model the process in this way (Fig. 7).

```
a.  Determination of variables

# Activity Machines in parallel with different processing times between machines
For m>=0                                  # Total machines to program
For n>=0                                  # Total machines to program

# Set of subscripts
Set Job:=1..n;                            #Job to be scheduled
Set Job:=1..m;                            # Machine to be scheduled

# Model parameter
Param P{Job, Machine}>=0 binary           # Processing time of job i in machine k

#Variables of model
var X{Job, Maquina}>=0 binary             # A job i is assigned or not to machine k
var Cmax>=0;                              #Makespan

b.  Determination of the Target Function

# Target Function

Minimice Makespan: Cmax;

c.  Determination of Restrictions

#Restrictions

Subject to S1 {j in Job}:                 # Each job must be carried out only on one machine

    Sum{ k in {Machine} } X [j,k]=1;

Subject to S2 {k  in Maquina }:           # The operation time of each machine cannot be greater than Cmax

    Sum{ j in {Job} } X [ j,k ]*P [j,k]<=Cmax
```

Fig. 7. Model programmed in AMPL

In this case, the analysis of the variables studied may be explained as follows topics:

- Pm: "m" Identical machines in parallel
- Prec: Precedence or mandatory restrictions between pairs of works
- Cmax: Max{C1, … ,Cn}: Maximizes the number of options (Xi),
- Minimizing completion time of all Cj jobs, also called programming length (makespan)

As a result of the assignment model, three work scenarios are established from the three parallel positions where the work is executed, taking into account work rhythms due to the expertise and maneuverability of the field teams. In that order of ideas it is established that the categories (simulated in the process) pretend to go to 70–80, 81–90 and 91–100% of the workers' abilities.

4 Discussion

Beyond these found results, it was possible to validate, with results from the Weibull probability distribution model (which is the most recommended model for the analysis of failure rates), that the values with which the workers' performance schemes are analyzed and indicated show failures (values outside of estimation parameters) in a time zone in which operations are executed in the hours in which the personnel must be in feeding or a break, especially at midday or lunch time; this leads to a restructuring of the planning tasks taking into account variables such as fatigue, tiredness or disconcentration in the process stages after 6 or 7 h of continuous work.

The critical path and the analysis of some modifications showed low probabilities that infer few possibilities of maneuver in a new internal planning of the tasks; that is to say the variables and contributing elements of the process inflexibility the activity, reason why there was the necessity to look for a strategy combined with the external logistic tasks through a model of linear Programming; Torres et al. (2018).

The linear programming optimization model proposes a Schedule to take advantage of operative capacities through the use of work levels according to individual performance: high medium and low, thus determining the use of warehouse management (doors in use) according to associated strategies, level of expertise and operator training.

The correlation between the studied variables represented in a linear fit model shows as a result that there is a high affinity between the parts: performance-wear-time, so we can approximate our future interpretations of the results of the linear model and the adjusted model as one and credible or reliable, which indicates that the correlation between the values of two scenarios is a fact. It depends on the interpretation of the phenomenon, and since correlation does not imply causality, it is pertinent to promote practices on other variables.

References

Aigües, P.M., et al.: Influencia de la fiabilidad humana en la evaluación del impacto en el riesgo de cambios de CLO (2014)

Mohamed, A.J., Udaiappan, M., Kumar Taisum, T., Srinivasan, S.: Reliability availability and maintainability study: a business perspective. Int. J. Perform. Eng. **9**(4) (2013)

Collier, Z.A., Hendrickson, D., Polmateer, T.L., Lambert, J.H.: Scenario analysis and PERT/CPM applied to strategic investment at an automated container port. ASCE-ASME J. Risk Uncertain. Eng. Syst. Part A Civ. Eng. **4**(3), 04018026 (2018)

Gonzalo San José, S.: IT RAMS: Modelización y simulación de la fiabilidad, disponibilidad y mantenibilidad de sistemas TIC: Estudio de caso práctico: clúster de aplicación (2016)

Montero, J.D.J.Á., Sarmiento, A.E.: La logística urbana, la ciudad logística y el ordenamiento territorial logístico. Revista RETO: Revista Especializada en Tecnologías Transversales de la Organización **4**(4), 21–40 (2017)

Nassar, M.M., Radwan, S.S., Elmasry, A.S.: The exponential modified weibull logistic distribution (EMWL). EPH Int. J. Math. Stat. **4**(1), 22–38 (2018). ISSN 2208-2212

Rahma, N.S.F.A., Notteboom, T., Rahmatdin, M.N., Othman, K.: Elección de puerto por parte de operadores de servicios de contenedores intrarregionales: una aplicación de técnicas de toma de decisiones a los servicios de línea entre los puertos de Malasia y otros puertos asiáticos (2019)

Pastor, J.J.C.: Apunte sobre el contrato de manipulación portuaria. Revista Boliviana de Derecho **24**, 424–443 (2017)

Predictive maintenance. Fault detection and diagnosis. Las Palmas de Gran Canaria University. Maintainability Module 3. Master's Degree in Reliability and Risk Engineering. Palmas de Gran Canaria, Spain (2007)

Rodriguez, S.B., Muñoz, A.H., Alvarez, M.E.S., González, J.A.T.: Las competencias y su relación con la competitividad en el área de la logística en el puerto de Altamira. Revista Boletín Redipe **4**(9), 110–122 (2017)

Santoyo, F.G., González, A.L.T., González, A.L.T.: Planning and scheduling techniques for efficient and effective project management. J. Manag. Sci. Res. **6**(10), 57–68 (2017)

Taha, H.A.: Investigación de operaciones. Pearson Educación, Editorial, Pearson (2004)

del Pino, T., María, J.: NPT 331: Fiabilidad: La distribución de Weibull (1994)

Torres, M.O., Valdés, P.M.F., Castillo, E.A.C.A.: Desempeño logistico y rentabilidad económica. Fundamentos teóricos y resultados prácticos/Logistic performance and economic probability. Theoretical grounds and practical results. Revista Economía y Desarrollo (Impresa) **149**(1) (2018)

Vergara, W.R.H., Teixeira, R.T., Yamanari, J.S.: Análise de risco em projetos de engenharia: uso do PERT/CPM com simulação. Exacta **15**(1), 75–88 (2017)

Bullwhip Effect in Perishables Supply Chain: Literature Review

Julián A. Durán[1](✉) and Ángel Ortiz[2]

[1] Industrial Engineer Program, Engineering Faculty, Universidad Autónoma de Bucaramanga, Bucaramanga, Colombia
jduran783@unab.edu.co

[2] Engineer and Industrial Production PhD Program, Universidad Politécnica de Valencia, Valencia, Spain
aortiz@cigip.upv.es

Abstract. The management of perishable products in the supply chain is taking on worldwide relevance due to the shortage of food that will be presented due to the increase in the world's population. The bullwhip effect is one of the justifications by which the level of inventories in the chain is increased and therefore also the waste of products. The present research seeks to know the reasons why this phenomenon occurs and the solutions that have been generated so far.

Resumen: La gestión de productos perecederos en la cadena de suministros esta tomando relevancia a nivel mundial por la escasez de alimentos que se presentará por el incremento de la población mundial. El efecto látigo es una de las justificaciones por las cuales se incrementa el nivel de inventarios en la cadena y por tanto tambien el desperdicio de productos. La presente investigación busca conocer las razones por las cuales se presenta éste fenómeno y las soluciones que se han generado hasta el momento.

Palabras Clave: Supply Chain · Perishables · Bullwhip effect · Demand · Forecast

1 Introduction

The Supply Chain Management can be approached from several perspectives, from the management of the suppliers, through the production and service systems and finally reaching the client's management. The goods that mobilized from the suppliers to the customer represent the money invested by each of the companies involved in the process; therefore the protection, location in the supply chain and quantity that is stored must be looked at in detail to avoid cost overruns and loss of merchandise. The bullwhip effect is nothing other than the increase of these goods along the chain for reasons of demand planning, distortion of the information transmitted between customer and supplier, fluctuation of sales prices and the missing game Between customer and supplier. Perishable goods have a particularity in this phenomenon, and it is the final disposal or disposal for most of them, while in the case of durable goods they should not necessarily be discarded, but some of them may have a second demand cycle. This research focuses on knowing the particularity of the supply chains of

M. T. Castañeda Galvis et al. (Eds.): ICIE 2019, LNMUINEN, pp. 159–164, 2021.
https://doi.org/10.1007/978-3-030-49370-7_17

perishable products, the possible variables that generate the bullwhip effect, and the different existing tools to reduce this phenomenon. In Sect. 1, there are the objectives of the investigation; Sect. 2 contemplates the methodology applied, in Sect. 3, the results showed so far, and in Sect. 5 are the conclusions.

2 Methodology

The research carried out is of an exploratory type, which was carried out through a review of available literature on the subject. The review began with the identification of the differences between traditional supply chains, which understood as those consisting of durable goods, and the supply chains of perishable products. It was necessary to identify variables that make up a supply chain and the particularities in each case. Secondly, those particularities were used to outline supply chains and then identify the problems that arise in each case.

The excess of inventory is a problem that identified in the documentary review, not only for the case of durable products but also for the case of perishable products, in each case its treatment is different but the result is a detriment of the value of the company through an increase in operating costs. This excess inventory increases as we look upstream of the final consumer, in some cases more than in others, so this phenomenon is known as the "bullwhip effect", and it was ultimately the final stage of this documentary review, a vision on the problems that the presence of the bullwhip effect entails in the supply chain of perishable products.

3 Results

The results seek in the first instance to contextualize the problem that exists around food losses and wastes, the possible causes and a view from Latin America and the Caribbean. Secondly, it is sought to understand how from the management of the supply chain, this phenomenon can be related to the bullwhip effect.

Food Loss and Waste

Researcher can talk about Food Loss and Waste (FLW) as the decrease in the mass of food for human consumption at any point in the production chain (United Nations Food and Agriculture Organization 2017), it is essential to indicate the difference between both concepts. Losses occur mainly during production, post-harvest, storage, and transportation; while waste occurs during distribution and consumption, directly related to the behavior of wholesalers and retailers, food sales services and consumers who decide to discard food that still has value.

About one third of the production of food intended for human consumption is lost or wasted worldwide, equivalent to approximately 1.3 billion tons per year (FAO 2011); food loss occurs at different stages of the supply chain and is caused by different forces, in addition this phenomenon differs between those countries with low income and medium and high income, while in countries with high-income food waste occurs in the production and post- harvest stage, in countries with medium and high income this is presented in the distribution and consumption stage (Food and Agriculture Organization of the United Nations 2018).

The following figure shows food losses and waste according to the location in the supply chain and the world region in which presented (Fig. 1).

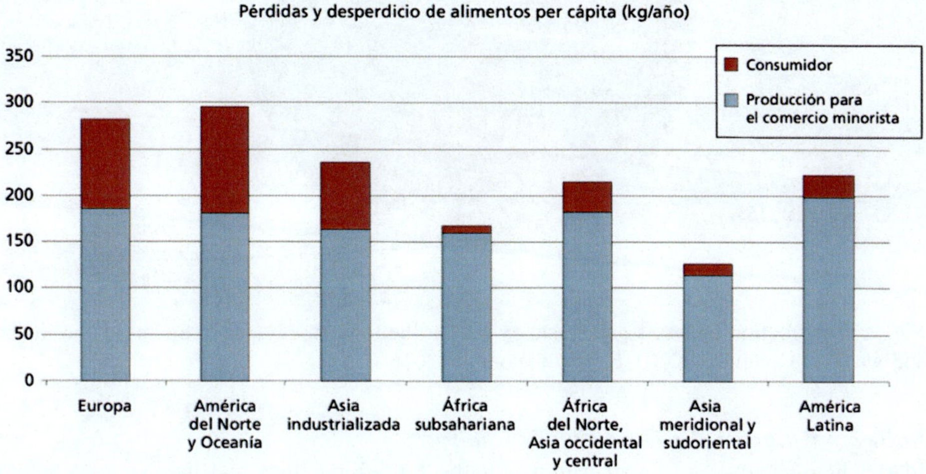

Fig. 1. Food losses and waste per capita in the consumption and pre-consumption phases in different regions. Source: Food losses and waste in the world, FAO. Dusseldorf-Germany, 2011

The previous picture establishes that in the high-income countries a large amount of food is lost throughout the supply chain, however a large amount is wasted in the final consumer, which means that the product discarded when they are still suitable for the consumption. On the other hand, countries with lower incomes show higher waste in the early stages of the food supply chain.

Faced with this problem, Latin American and Caribbean countries have emerged that apply regulations to reduce the level of food waste, for the Colombian case there is bill # 164 on April 2016, which creates the National Food Program against food waste (PANDA), which seeks to implement measures and obligations for food producers, processors, distributors, and consumers, in order to reduce the percentage of food discarded in the country.

The Food and Agriculture Organization FAO designed the Mass Flow Method MAS in order to account for food losses and waste at each stage of the Food Supply Chain, in 2016 the National Planning Department Colombia conducted the first FLW study based on this methodology. The result of the study showed that the available supply of food is 28.5 million tons, and 9.76 million tons are lost and wasted, equivalent to 34% of the total, in a few words one in three tons of food is lost or wasted (National Planning Department 2016). Of the whole food that is lost or wasted, 64% correspond to stages of production, post-harvest, storage, and industrial processing; the remaining 36% corresponds to the stages of distribution and retail, and household consumption (Fig. 2).

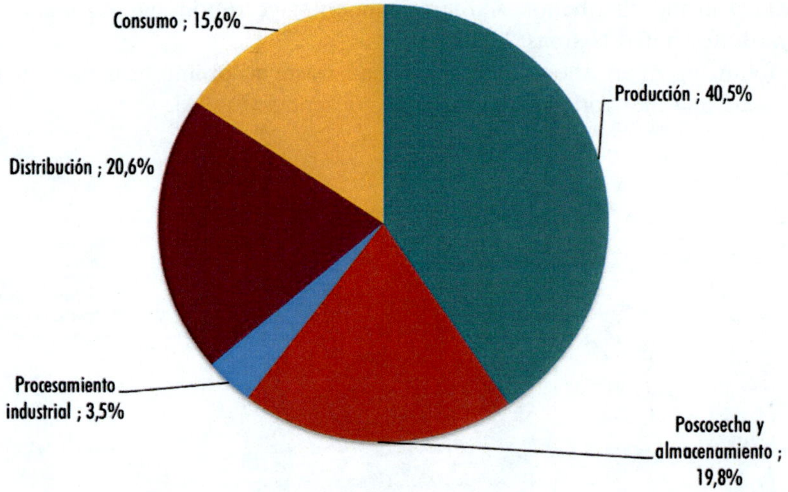

Fig. 2. Distribution of losses and waste by link in the food supply chain. Source: Food losses and waste in Colombia, DNP, Bogotá-Colombia; 2016

Bullwhip Effect on the Food Supply Chain

In the literary review, it had found that the bullwhip effect had its first conceptual example with Jay Forrester in his book Industrial Dynamics in 1961, when he wielded the effects that the distortion of information among the actors of a group may have on supply chains. Productive chain. In more recent years (Lee et al. 1997) he was able to establish four reasons that may be generating the bullwhip effect in supply chains: price fluctuations, updating customer demand, production batches and the game of rational missing. Zarandi et al. (2008) have found that also the delivery time of the supplier, the type of product whether perishable or not, the variability of the purchase orders, and the level of interaction between substitute and complementary products can increase or decrease the variability of demand.

The Bullwhip effect can defined as the increase in the variability of the demand upstream of the final customer, which is the result of the distortion in the information transmitted between the companies providing the final product (Padmanabhan et al. 1997). The way to measure the bullwhip effect is the division between the variation of the order orders that it transmits and those that it receives:

$$Bw = \frac{Var_{(}}{Var_{)}}$$

Researchers have developed tools to reduce this variation in demand in supply chains, share information among the actors in the chain to have greater certainty of the final demand of the client, model of system dynamics to understand the accumulation of inventory and variables that must be managed to reduce the effect, inventory management to avoid their increase, specialized forecasting techniques such as neural networks, fuzzy logic, multi-agent systems, ARIMA model, among others, hybrid forecasting models used lately. Which are generating better results than traditional models.

When reviewing a variable similar to the bullwhip effect, the inventory; found that 4% of the investigations are around inventory management (Siddh et al. 2017). This position can be seen in a different way when compared to the use of new practices such as Value Stream Mapping to reduce food losses in the supply chain (De Steur et al. 2016), the which invites us to review the consequences of the bullwhip on the loss of food in supply chains.

Supply chain managers indicate that food waste may be due to factors such as shared information and forecasts and order orders among others (Mena et al. 2011), there are two of the reasons raised by (Lee et al. 1997) why found the bullwhip effect in supply chains. In the case of the demand for perishable products, it should be in mind that the discounts have a close relationship with other decisions in the supply chain, such is the case of the purchase order and the space available for storage (Yang et al. 2017).

For perishable products, management of the bullwhip effect varies depending on the expiration level of the products distributed. In the expiration level, it was possible to identify what factors like accuracy the forecast, the delivery time of the provider, and the information that is shared play an essential role in reducing the bullwhip effect.

4 Discussion

The optimization of supply chains can generate competitive advantages from the point of costs and the level of service. The bullwhip effect generates an increase in the costs of the chain, as this increases the levels of inventory, operational overruns and lack of products with high turnover, investigating their implications allows reducing their impacts and improving forecasting and inventory policies. Given the global importance for the effects of food shortages that the planet may suffer in the coming years, researchers have taken on the task of further investigating food waste that occurs in developed countries, while in others it happens otherwise; that is why the existence of the bullwhip effect in these types of supply chains is of relative importance for different organizations and ONG's that fight against social inequality.

The relationship between the loss and waste of food with the supply chain is an invitation to propose solutions that improve the quality of life of society. This research aimed at generating solutions that reduce this phenomenon, improving forecasting systems or working together to be sure of the final demand of the customer, can avoid the most significant food waste that occurs daily in our country.

References

Asia Pacific Economic Cooperation - Agriculture Technical Cooperation Working Group: APEC Survey Report on Feasible Solutions for Food Loss and Waste Reduction. APEC, Taipei (2018)

De Steur, H., Wesana, J., Dora, M.K., Pearce, D., Gellynck, X.: Applying value stream mapping to reduce food losses and wastes in supply chains: a systematic review. Waste Manag. **58**, 359–368 (2016)

Departamento Nacional de Planeación: Pérdida y desperdicio de alimentos en Colombia. DNP, Bogotá (2016)

FAO: Pérdidas y desperdicios de alimentos en el mundo. Dusseldorf. ONU, Alemania (2011)

Fazel Zarandi, M., Pourakbar, M., Turksen, I.: A Fuzzy agent-based model for reduction of bullwhip effect in supply chain systems. Expert Syst. Appl. **34**, 1680–1691 (2008)

Food and Agriculture Organization of the United Nations: Food loss and waste and the right to adequate food: Making the connection. ONU, Roma (2018)

Lee, H.L., Padmanabhan, V., Whang, S.: The bullwhip effect in supply chains. Sloan Manag. Rev. **38**, 93–102 (1997)

Mena, C., Adenso-Diaz, B., Yurt, O.: The causes of food waste in the supplier– retailer interface: evidences from the UK and Spain. Resour. Conserv. Recycl. **55**(6), 648–658 (2011)

Metters, R.: Quantifying the bullwhip effect in supply chains. J. Oper. Manag. **15**, 89–100 (1997)

Organización de las Naciones Unidas para la Alimentación y la Agricultura: Pérdidas y desperdicios de alimentos en América Latina y el Caribe. ONU, Nueva York (2017)

Padmanabhan, V., Lee, H., Whang, S.: Information distortion in a supply chain: the bullwhip effect. Manag. Sci. **43**(4), 546–558 (1997)

Siddh, M.M., Soni, G., Jain, R., Sharma, M.K., Yadav, V.: Agri-fresh food supply chain quality (AFSCQ): a literature review. Ind. Manag. Data Syst. **117**(9), 2015–2044 (2017)

Yang, S., Xiao, Y., Kuo, Y.-H.: The supply chain design for perishable food with stochastic demand. Sustainability (Switzerland) **9**(7), 1195 (2017)

Zarandi, M.H.F., Pourakbar, M., Turksen, I.B.: A fuzzy agent-based model for reduction of bullwhip effect in supply chain systems. Expert Syst. Appl. **34**(3), 1680–1691 (2008)

Supply Chain Integration Towards Global Value Networks - Review

Woody Figueroa Peinado[1], Benjamín Pinzón H.[2], José M. Díaz P.[1](✉), and José D. Gómez M.[2]

[1] Facultad de Ciencias Empresariales, UNIMINUTO UVD, Bogotá DC, Colombia
{woody.figueroa, jose.diazp}@uniminuto.edu, woody_figueroa@yahoo.com, ingmartind@gmail.com
[2] Facultad de Ingeniería, UNAD, Bogotá DC, Colombia
{Benjamín.pinzon, josed.gomez}@unad.edu.co, bpinzonh@gmail.com, dangomezl9@hotmail.com

Abstract. The global environment enhances new opportunities for countries creating a broader framework in terms of economy, trade, social development, technologies and knowledge. In that way, enterprises have challenges not only for domestic markets but also for their insertion through international markets, where competitiveness is crucial. Therefore, it couldn't be understood as an individual effort from organizations. But rather as the capacity to generate integration of supply chains from the local and global environment, creating, strengthening and consolidating global value networks.

Keywords: Industrial Management · International trade · Supply and demand · Transport planning

1 Introduction

This article shows the evolution from supply chains to supply networks and their consolidation as global value networks, analyzing promotion factors that make possible the creation of supply networks into generation value, the impact of relationship management and logistics costs in the integration processes. Similarly, this article takes into account policies that influencing in the strengthening and sustainability of global value networks considering the profile of high, medium and low-income countries.

2 Methodology

Methodology for this research is based on secondary sources, searching keywords on Scopus and segmenting authors in order or relevance and considering international journals such as: PROQUEST, EMERALD, EBSCO, IEEE, ACM, JSTOR, SCIENCE DIRECT, and others, including papers from international organizations, like: World Bank, World Trade Organization and World Economic Forum, among others. This research is driven through the evolution of the supply chain, the concept of value generation and global value networks. Subsequently, a keyword validation is done on UNESCO Thesaurus dictionary.

M. T. Castañeda Galvis et al. (Eds.): ICIE 2019, LNMUINEN, pp. 165–173, 2021.
https://doi.org/10.1007/978-3-030-49370-7_18

3 Results

This bibliometric review is focused in four main topics related with integration processes of the supply chain: the evolution from supply chain to supply networks, the generation of value, relationship management and policies aimed at consolidating global value chains.

3.1 From Supply Chain towards Supply Networks

From business environment, the evolution of logistics to what is now known as Supply Chain dates back to the end of the Second World War, at that time the concept was understood in terms of physical distribution, and logistics was not considered as a discipline (Ballou, 1978) cited in: Habib (2011). However, as competition increased, manufacturing companies should focus on ensuring operational efficiencies, thus consolidating logistics as a discipline, (Heskett et al., 2011), cited in: Habib., (2011), companies incorporate logistics management into their corporate strategy, at all levels: strategic, tactical and operational, being an individual effort to respond to the market.

The term Supply Chain management SCM is accredited to consultants Oliver & Webber (1982), which is subsequently introduced into the academy by Ellram & Cooper (1990); Jones & Riley (1987), who unify procurement, operations and distribution into the discipline of supply chain management. Cited in: Carter, Rogers, and Choi. (2015). As the discipline acquires a degree of maturity, new theories about supply chain management emerge. This is how Ballou (1992) incorporates the concept of logistics management in business, as an explanatory and support component for supply chain management; consequently, Bowersox et al. (2002), include aspects related to the interaction with customers, and strategic approaches to decision making as an integral and synergistic way "… transport management, storage, material handling and its integration in the logistics networks of the organization…." Cited in: Servicio Nacional de Aprendizaje, SENA (2014).

In this way, resource planning goes from being an individual exercise of companies to being oriented to the synchronization of processes from the operational and tactical component between companies which is known as extended company resource planning. In turn, the changes in the organizational structure of companies go from the functional or departmental structure towards the structure of processes, generating matrix arrangements that allow the formation of internal networks in the organization that migrate towards the generation of interactions and communication connections with suppliers and customers. As a result, it turns in a new philosophy of supply chain management. (Marujo, 2006), cited in: Feyissa, Sharma, Lai (2019). In this way, resource planning is oriented towards the collaboration of the actors in the supply chain, integrating the strategic, tactical and operational component, what is known as federated planning, where cooperation and collaboration require a fundamental component for each of the actors involved, which is trust.

The need to compete in the market, to establish differentiation factors, to respond more effectively to the expectations of customers and consumers, make companies address the need to generate integrations that allow them to configure supply networks with reliable business partners. For this it is necessary to define what process is likely to integrate, who are the suppliers and customers that can become business partners, and what level of integration is likely to be generated. Those insights make possible to define a Structural network arrangement and how to manage all its components (Fig. 1).

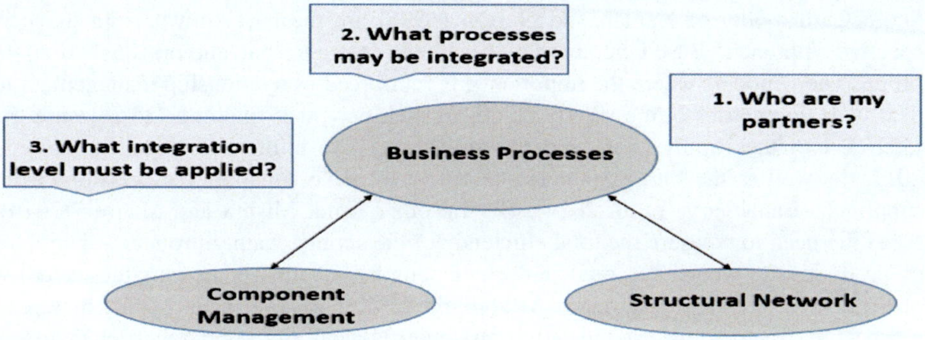

Fig. 1. Supply chain networks integration. Adapted from: Cooper, M; Lambert, P. J.D (2004)

3.2 Relationship Management in the Integration Processes

Relationship management in supply networks allows and shapes integration processes, migrating from the concept of supply chain to networks and then towards global value networks from the perspective of knowledge, metrics and innovation. Cerchione and Esposito (2016) report a compendium on the evolution of relationships in supply chain management, from direct supplier-customer relations (years 70s), cost-based relationships (years 80s), strategic alliance relationships (years 90s), globalization relations (2000 year) and relations based on the circular economy (up 2000 year) and the effect of digital disruption with industry 4.0, where business models in supply chains are refocused towards the culture of collaboration in both knowledge and information, processes tailored to the supply chain and service-oriented technology, infrastructure and security. World Economic Forum (2016).

Based on relationship management, knowledge management practices are identified. Cerchione and Esposito (2016) make a systematic review, analyzing different authors and considering three perspectives as follow: network relationships, knowledge management process and barriers. Network relationships include internal and external relationships between interdependent companies and business units that support forward and backward flows, from the initial provider to the end customer, in order to achieve customer satisfaction. (Boyer & Stock, 2009). To make knowledge management possible, trust and confidence facilitate an integration process from customers until suppliers at first, second and third level, (Li et al. 2011), and that formal and informal mechanisms must be established for knowledge transference by means of relationships management (Zhang & Zhou, 2013). Other factors such as: electronic commerce, integration knowledge of capacities to assume risk factors of environment, make possible strengthening networks (Liu, 2015, 2016). However, knowledge management in the supply chain still has barriers to face from the point of view of strategic, organizational and cultural relations because it is not integrated with the business processes of the supply network (Patil & Kant, 2014).

Balfaqih et al. (2016), make a review about the literature on measurement systems from 1998 to 2015, identifying three approaches to them based on different authors as follow: perspective, processes and hierarchy. The SCOR model (1996) and the Balance Score Card model of Kaplan and Norton (1996) are the best known from the perspective approach. Based on these two models, more recent authors have tried to propose new models where the importance is reinforced of relationship management to deal with the metrics in the supply chain; in such way, it is important to measure the total cost of the supply chain and the ability to share information (Hall & Saygin, 2012), as well as the Intra-organizational effect to make collaborative decisions from suppliers; (Battacharya et al. 2014). On the other hand, Mishra and Sharma (2014), focus the need to measure the total efficiency of the supply chain, introducing variables of productivity, working capital and chain length, but also other variables such as quality, flexibility and integration (Balfaquih & Yunus, 2014); and from hierarchy perspective, metrics are established at strategic, tactical and operational level, incorporating performance indicators in the skill management of the supply chain (Lauras et al. 2011), as well as the ability to provide inter-operational solutions (inter-companies) for a particular environment (Galasso, 2014).

From the point of view of innovation processes in the supply chain, Wong and Ngai (2019) conduct a literature review from 1999 to 2016. A trend to generate innovation is identified from the collaboration in the integration of the actors: supplier + manufacturer + distributor; When these actors work together to innovate, they generate greater economic results. (Mac Carthy et al. 2016) who refer to the integration of producers and suppliers allows the transformation of business models from supply while generating competitive advantages. In this way, innovation in the supply chain is an exercise that must be carried out at different levels, it is not simply an individual effort of the organization; hence the importance of relationship management and the collaboration of involved actors.

The management of relationships in the supply chain is a vital element that enables integration processes for the consolidation of efficient, flexible and innovative networks, for which knowledge management is a shared element that must be assumed by the actors involved in the management of interconnected processes and flows.

3.3 The Concept of Value in Supply Networks

The concept of the value chain is introduced by Porter (1980), understanding the organization as a chain of processes that integrate it generating value; in order words, deals with suppliers, manufacturers, distributors and retailers that ensure product delivery to the final customer. This way, the exchanges between each of the actors occur in the transactional field or trade-offs between the companies and in this way, it becomes a competitive advantage.

According to Acevedo and Gómez (2016), value network management, is "…an integrated system of activities including : planning, programming and a capable structure of carrying out and controlling a knowledge model that impacts the well-being of consumers, by means of efficient resource flows as a result of integration that makes possible add value from suppliers to final clients. Thus, according to these authors, "in the coming years, the concept of value network will prevail in the international field as a generalizer and integrator of the concept of supply chain and the way of doing business in a competitive way, establishing standards of behavior for the survival of organizations…"

On the other hand, digital disruption with industry 4.0 enables new frontiers in terms of generating value in supply networks. This is the case of Chernova et al. (2020), referring to the transformation of value networks for the retail sector in Russia, where the structure of service value is implemented for a large commercial network, making an ecosystem. Customers may carry out all services through a mobile application. This is a real digital transformation. That way, modern manufacturers offer a wide range of customer services applying big data, such as: forecasts, marketing intelligence, digitalization and innovative strategies that allow clients to participate in the joint creation of customer value to meet their needs, providing customer value quickly and geographically as they needed.

Another impact of the digital era on value generation in the supply networks is the role of service providers. It is the case mentioned by Often (2018), where Blume company works for providing technological services that allow the digital transformation of supply networks, whose purpose is to unite all the elements of the global supply network, from retailers and manufacturers to logistics service providers, to boost innovation and growth, leveling the cost and efficiency to offer a new added value.

3.4 The Strengthening of Global Value Networks

The change in global production and supply schemes migrates from the simple transactional exercise of exports and imports to the generation of high value of products and services, based on the configuration of possible supply chains. According with World Bank (2017), transactions that are made in the supply network are not simply products but also the value that is being exchanged in each of the stages (Diakantoni et al. 2017). This is how the consolidation of global value networks arises with the structure of production models and the number of steps or stages that make up the supply chain until reaching the consumer or end user. In this sense, productive activities begin overcoming local barriers, which means that production schemes start to have a global approach. In this sense, the generation of value networks is outlined as value is added to the product in each of the stages of the supply chain (Degain, Meng and Wang 2017).

Wang et al. (2017), propose four ways to consolidate global value networks: domestic production with added value generation in domestic markets; export of finished products, as a traditional trade; global simple value networks (off-shore operations of semi-finished products with re-entry to the country of origin) and complex global value networks (off-shore operations of finished products) (Fig. 2).

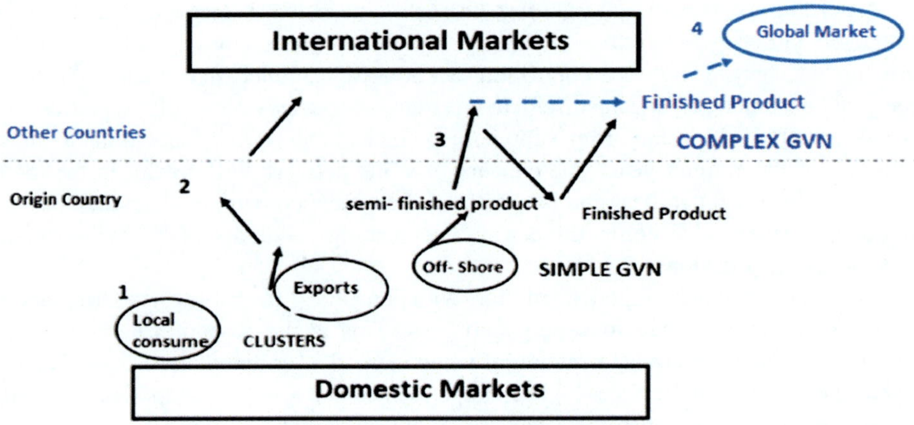

Fig. 2. Typology of global value networks

Based on Wang et al. (2017) Off-shore operations involve the generation of alliances or partnerships for generating competitive advantages to concentrate on a specific production process or task, which makes it easier to integrate into the global economy (Kowalski et al. 2015). When manufacturing processes are geographically segmented and organized as a global production network, a cascading effect is generated in the supply network that at the end must be assumed by the end consumer. Koopman, Wang and Wei (2014), cited in World Bank (2017), state the importance of generating institutional policies that promote economic benefits in terms of tariffs, customs costs for the development of free zones and international trade, with the logistics cost being a determining factor. Those nations whose export of added value has greater weight in their level of participation in global value networks, require a greater economic and logistic development than those whose imports are more notable (Martí & Puerta, 2019), cited in: World Bank (2017).

In this sense, Diakantoni and others (2017) have correlated the Page Rank indicator with the logistic performance index, to assess the costs incurred for commercial exchange and measure the efficiency of supply networks comparing countries, such that small economies or low-income countries have a higher commercial exchange cost and therefore it is more difficult to consolidate global value networks. For this, the World Bank (2017) outlines that it is necessary to develop policies and indicators that allow the generation, consolidation and strengthening of global value networks, according to the profile of the countries: low, medium and high income (Table 1).

Table 1. Policies towards global value networks

Stage	Beginning	Strengthening	Consolidation
Country Profile	Low Income	Middle Income	High Income
Actions to develop	- Foreign Direct Investment - Insertion of domestic companies in global networks. - Efficient uses of resources.	-Participation in networks. - Reinforced capacities in domestic markets. - Access to capitals and commercial agreements.	-Sustainability and transformation. -Stable policies to develop networks. - Skills performance - Social promoting
Indicators	Logistics Performance Index Doing Business Labor Costs	Research and development Certification ISO	Social security Wage statistics Labor Standards

Based on Taglioni y Winkler (2016) for World Bank. Making Global value chains for development

4 Discussion

Global markets make economies interconnected, competitiveness and sustainability are emerging from the ability to generate global value networks, either from domestic markets with the participation of companies for the development of clusters, as well as the insertion into the internationalization processes in some of the stages of the supply network. However, to make it possible, it is necessary that the organization is prepared from its corporate strategy, integrating its own processes within the organization and once it is prepared, generate the channels that allow its integration with the actors involved in its own supply chain, customers, suppliers of materials, logistics, financial and, technological service providers and all those who can participate to consolidate supply networks.

The integration processes represent an important challenge from the costs and the generation of value; those countries that have a high volume of export require a greater study of global value network, due to high imports and exports volumes, transactions and flows generated in the networks that make up, compared to those countries that focus on their domestic market. Whether global value networks are developed at the local or international market, relationship management is required from collaboration with the actors involved, which contribute to the efficiency of processes, technological incorporation, adaptability and flexibility, for which it is necessary to promote a shared knowledge management in the supply network aimed at innovation and sustainability, assuming the risks and challenges of the digital age.

Acknowledgments. Author thank sponsors who make possible this chapter book, by means of funding support from Universidad Pontificia Bolivariana (UPB) seccional Bucaramanga and technical support from Corporacion Universitaria Minuto de Dios (UNIMINUTO) sede Virtual y Distancia Bogotá Calle 80 and Universidad Nacional Abierta y a Distancia (UNAD) sede José Acevedo y Gomez Bogotá – Cundinamarca, Colombia.

References

Acevedo, J.A.Y., Gómez, M.I.: Modelo de Referencia de la Red de Valor en Latinoamérica. Universidad Tecnológica de la Habana: Recuperado de (2016) https://www.researchgate.net/publication/311992835_Modelo_de_Referencia_de_la_Red_de_Valor_en_Latinoamerica/link/5867c85608ae8fce4915c5a9/download

Balfaqih, H., et al.: Review of supply chain performance measurement systems: 1998-2015. Comput. Ind. **82**, 135–150 (2016). https://sciencedirect.ezproxy.uniminuto.edu/science/article/pii/S0166361516301166

Carter, C.R., Roger, D.S., Choi, T.Y., et al.: Toward the theory of supply chain. J. Supply Chain Manage. **51**(2), 89–97 (2015). https://search-proquest-com.ezproxy.uniminuto.edu/docview/1678722040?accountid=48797

Cerchione, R., Esposito, E., Choi, T.Y., et al.: A systematic review of supply chain knowledge management research: State of the art and research opportunities. Int. J. Prod. Econ. **182**, 276–292 (2016). https://doi.org/10.1016/j.ijpe.2016.09.006

Chernova, D.V., Sharafutdinova, N.S., Nurtdinov, I.I., Valeeva, Y.S., Kuzmina, L.I.: The transformation of the customer value of retail network services under digitalization. In: Ashmarina, S.I., Vochozka, M., Mantulenko, V.V. (eds.) ISCDTE 2019. LNNS, vol. 84, pp. 252–260. Springer, Cham (2020). https://doi.org/10.1007/978-3-030-27015-5_31

Diakantoni, A., Escaith, H., Roberts, M., Verbeet, T.: Accumulating Trade Costs and competitiveness in Global Value Chains. World Trade Organization (WTO) Working Paper Economic Research and Statistics Division (ERSD) 2017–2, WTO (2017) https://www.wto.org/english/res_e/reser_e/ersd201702_e.pdf

Degain, M.W.: Recent trends in Global Value Chains. World Bank, Measuring and Analyzing the impact of GVC on economic development (2017). http://www.wto.org

Feyissa, T.T.R., Sharma, R.K., Lai, K.K.: The impact of the core company's strategy on the dimensions of supply chain integration. Int. J. Logist. Manag. **30**(1), 231–260 (2019). https://doi.org/10.1108/IJLM-03-2017-0080

Kowalski, P., et al.: Participation of Developing Countries in Global Value Chains: Implications for Trade and Trade-Related Policies OECD Trade Policy Papers, No. 179, OECD Publishing, Paris (2015)

MacCarthy, B.L., Blome, C., Olhager, J., Jagjit, S.S., Zhao, X.: Supply chain evolution theory, concepts and science. Int. J. Oper. Prod. Manage. **36**(12), 1696–1718 (2016). https://doi.org/10.1108/ijopm-02-2016-0080

Habib, M.: Supply Chain Management: Theory and Evolution (2011). www.intechopen.com

Often, L.: Blume Global presenta una era digital de crecimiento, innovación y creación de valor para las cadenas de suministro. In: Asian & European Business Collection. United States, New York (2018). https://search-proquest-com.ezproxy.uniminuto.edu/asianeuropeanbusiness/docview/2105027560/52863849E77F4CC2PQ/24?accountid=48797

Pedersen, T., Devinney, T.M., Tihanyi, L.: Breaking up the Global Value Chain: Opportunities and Consequences. Bingley, UK: Emerald Publishing Limited (2017)

Pietrobelli, C.Y., Staritz, C.: Cadenas Globales de Valor y Desarrollo Económico. Desarrollo Económico **56**(220), 371–391 (2017). https://www.jstor.org/stable/44736001

Porter, M.: Estrategia y ventaja competitiva. (1980). https://ebookcentral.proquest.com/lib/bibliouniminutosp/detail.action?docID=4824578

Servicio Nacional de Aprendizaje, SENA. Caracterización de la Logistica en Colombia (2014). http://wcl.com.co/wp-content/uploads/2016/01/LOGISTICA-EN-COLOMBIA-2014.pdf

Taglioni, D., Winkler, D.: Making Global value chains for development (2016) https://openknowledge.worldbank.org/handle/10986/24426

Wang, Z., Wei, S.-J., Yu, X., Zhu, K.: Measures of Participation in Global Value Chains and Global Business Cycles. Working Paper 23222, National Bureau of Economic Research, Cambridge (2017) https://www.nber.org/papers/w23261.pdf

Wong, D.T., Ngai, E.W.: Critical review of supply chain innovation research (1999–2016). Ind. Market. Manage. **82**, 158–187 (2019). https://doi.org/10.1016/j.indmarman.2019.01.017

World Economic Forum, and Accenture. Digital Enterprise. Digital Transformation of industries (2016) http://reports.weforum.org/digital-transformation/wp-content/blogs.dir/94/mp/files/pages/files/digital-enterprise-narrative-final-january-2016.pdf

World Bank Group et Al. Global Value Chain Development Report. Measuring and Analyzing the impact of GVCs on economic development (2017) https://www.wto.org/english/res_e/publications_e/gvcd_report_17_e.htm

Growth Curves and Evolution of Land Transportation in Colombia 2001 to 2017

Wilson A. Pinzón R.[1](✉), Diego E. K. Rodríguez[2], and José M. Díaz P.[2]

[1] Facultad Tecnológica, Universidad Distrital "Francisco José de Caldas", Bogotá, Colombia
wapinzon@udistrital.edu.co

[2] Facultad Ciencias Empresariales, Corporación Universitaria Minuto de Dios, Bogotá, Colombia
drodr191@uniminuto.edu.co

Abstract. This is where the flow of cargo and land transport in Colombia from 2001 to 2017 is characterized. Logistic curve fitting and Gompertz models are used. Simultaneous cyclic behaviors are identified using wavelet models. Limits were found to the growth of 4 of the 6 modes of transport.

Keywords: Freight and Passenger Transportation · Land and Air Transportation · Colombia · Gompertz

1 Introduction

This text describes the simultaneous evolution of six types of land transport in Colombia: the flow of automobiles, two-axle buses, three-axle and four-axle trucks, five-axle trucks, six-axle trucks and integrated mass transport systems. To describe them, it uses time curve fit, logistics and gompertz models, among others, to identify growth and evolution potentials. It also uses component identification with simultaneous cycling using Wavelet models.

2 Methodology

At the beginning, the literature is reviewed and the available public information is obtained from the flow of land transportation grouped by the aggregate type of flow in each month from 2001 to 2017 for: automobiles, two-axle buses, three-axle and four-axle trucks, five-axle trucks, six-axle trucks, and seven integrated mass transportation systems. The growth curve is then identified and the flow evolution parameters are estimated. Third, the simultaneity in the behavior of several modalities is identified using techniques and models of signal and wave analysis such as spectral coherence and the Wavelet theory to predict the evolution of the land transport flow in Colombia.

M. T. Castañeda Galvis et al. (Eds.): ICIE 2019, LNMUINEN, pp. 174–179, 2021.
https://doi.org/10.1007/978-3-030-49370-7_19

3 Results

(i) State of the art review

François and Manchin find the relationship between infrastructure and institutional quality in international trade. They conclude that import and export trade depends on institutional access and on the adequate development of transport and telecommunications technologies, aspects that diminish resistance to the flow of trade (Francois and Manchin 2013), (Ueki 2015).

Ganni relates exports directly with the logistics performance index, in each of six aspects measured; with respect to imports only two of the six factors are relevant (Gani 2017), so he concludes that the quality of logistics affects the international trade of the countries, and from there he derives what improvements in logistics service lead to improvements in international trade, this on the 5 variables of the LPI, Logistics Performance Index, for 60 countries for the years 2007, 2010, 2012 and 2014. Given that market movement depends on the infrastructure used, and this changes within Latin American countries; German and its working group compares infrastructure across multiple countries using Data Enveloping Analysis, DEA (Suarez-German, Serebrisky, and Perelman 2019) and identifies different categories or groups in which they are classified with respect to 200 countries during the period 1996 to 2016.

Since logistical performance is associated with the availability and use of information technology, and in the case of Colombia, the effect of data availability on service provision is studied, and data availability is identified as a necessary but not sufficient condition for the provision of such services, within the customer service orientation (Grimaldi et al. 2019).

On the other hand, it is noted that the positioning strategies of logistic transporters affect the possibilities of trade between nations since they affect the commercial possibilities of individual exporters, since the private choice of geographic positioning of transporters affects the strategic positioning of their clients in globalized markets, according to Ghiara and Caminati (Ghiara and Caminati 2017). This situation changes for different means of transport and different markets and countries, such as passenger air transport in Colombia (Olariaga et al. 2018).

(ii) Data and information available

The Mystery of Transportation presents in the open data portal of the government of Colombia, and on its website the movement data on highways classified into the categories of: automobiles, two-axle buses, three-axle and four-axle trucks, five-axle trucks, six-axle trucks and integrated mass transit systems, available and intercomparable data from January 2000 to June 2017 or 198 monthly data for 6 variables and for total records and subtotal freight transport. This data is used in the "Curve Expert" software to determine the shape of curves that fit this data (Illustration 1, 2, and 3).

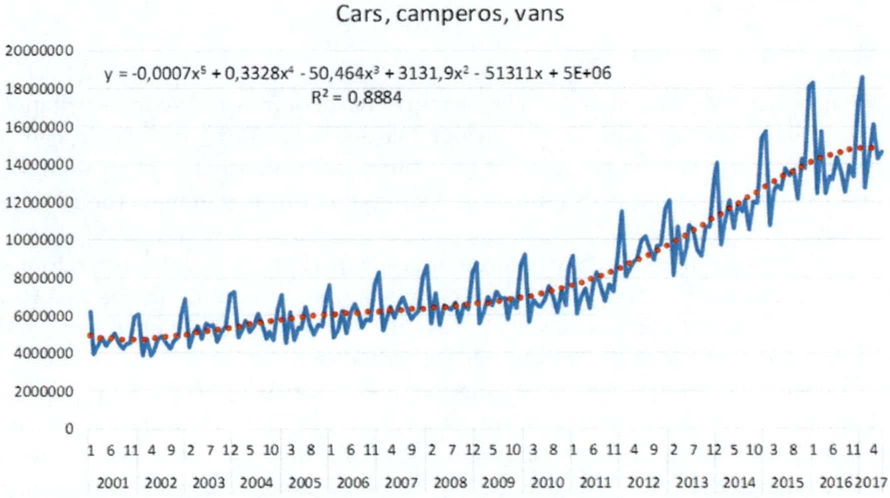

Illustration 1. Number of Monthly Trips, 2001 to 2017, Adjustment data and polynomial. Camperos and vans. Source Min Transporte, Colombia.

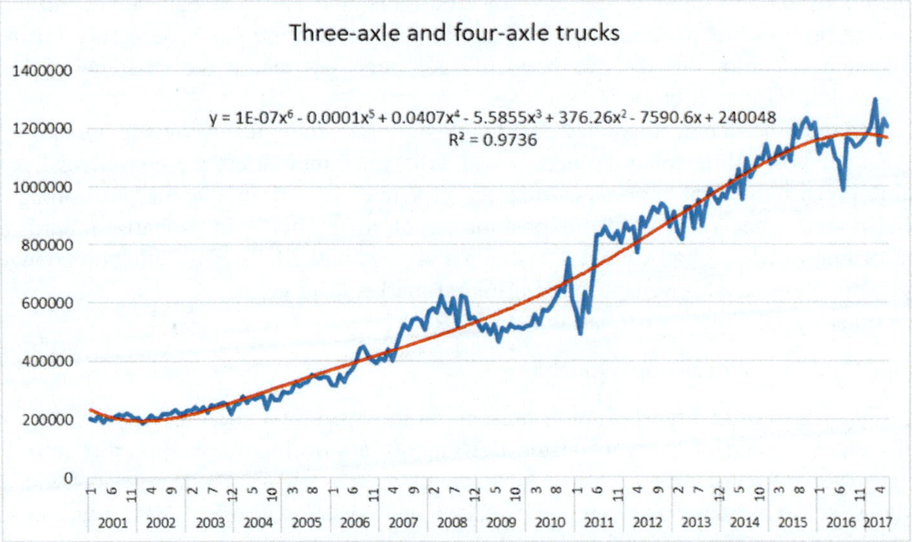

Illustration 2. Number of Monthly Trips, 2001 to 2017, Adjustment data and polynomial. Three-axle and four-axle trucks. Source Min Transporte, Colombia.

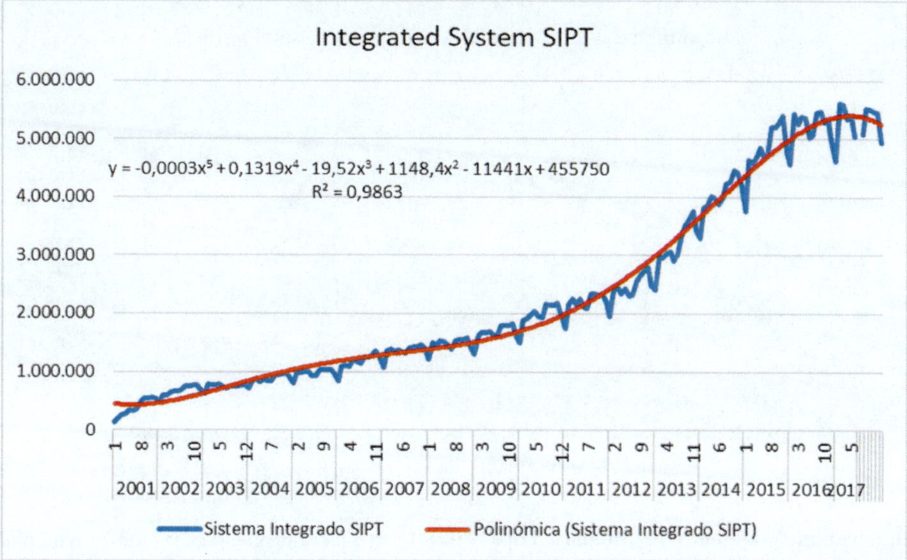

Illustration 3. Number of Monthly Trips, 2001 to 2017, Adjustment data and polynomial. Integrated Mass Transport System, SITP. Source Min Transporte, Colombia.

The polynomial adjustments are in the illustrations, in the first are the number of journeys by private motor vehicles vans and camperos; in the second the three and four-axle trucks and in the third the total for integrated mass transport transport.

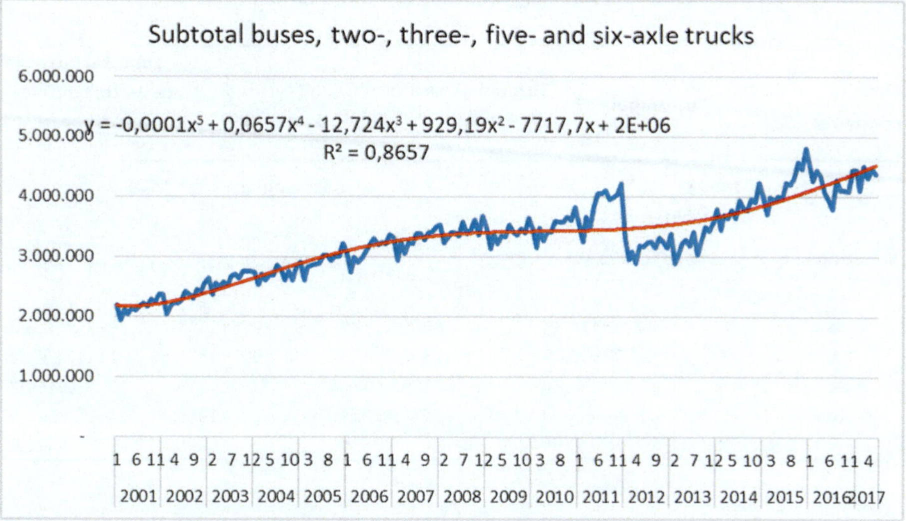

Illustration 4. Number of Monthly Trips, 2001 to 2017, Adjustment data and polynomial. Subtotals two-, three-axle trucks and buses. Source Min Transporte, Colombia.

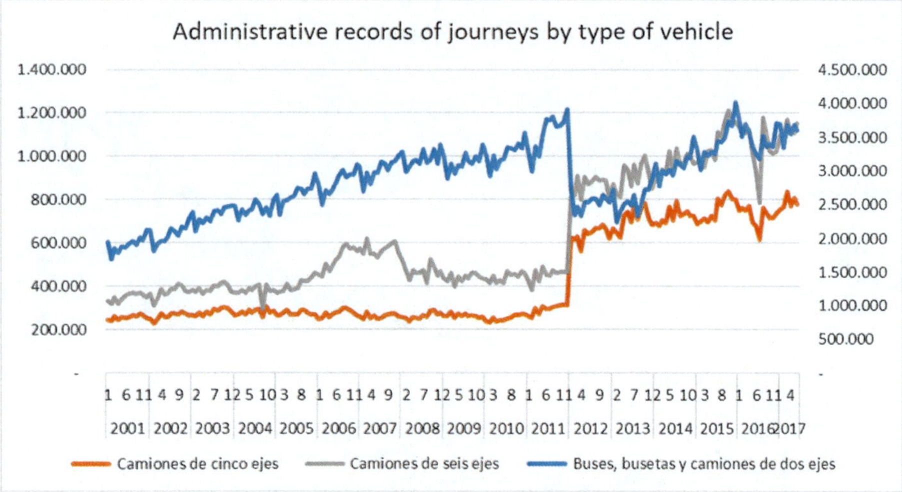

Illustration 5. Number of Monthly Trips, 2001 to 2017, Adjustment data and polynomial. Various categories. Source Min Transporte, Colombia.

The fourth illustration includes a subtotal with different classifications Illustration 4, since as can be seen in Illustration 5, the administrative records for these categories show disruptions in the number of records based on change in administrative criteria.

The equations describing the datasets are in Table 1.

Table 1. Modeling equations of transport datasets.

Data camperos, vans	Automobiles,	Three-axle and four-axle trucks	Integrated System SIPT	Subtotal buses, two-, three-, five- and six-axle
	trucks	Model	Sinusoidal:	MMF :
Equation	Sinusoidal: y=a+b*cos(cx+d)	Sinusoidal: y=(a*b+c*x^d)/(b+x^d)	y=a+b*cos(cx+d)	y=a+b*cos(cx+d)
a =	57340374,35	160691,2199	291645798	5635715,666
b =	52397590,75	22343,27868	290993381	11111239,86
c =	0,003734824	8765939,584	1052,9856	0,001316085
d =	3,04754758	1,530112011	3,11955	4,422859635
R^2	0,9294	0,9833	0,945	0,92347

4 Discussion

The available data are not homogeneous for all categories as they present changes in the order of magnitude at specific times at the beginning of 2012. For this reason, the data are grouped into four categories: cars, vans and campers; three- and four-axle trucks; integrated passenger transport systems; buses, two-, four- and five-axle trucks; and buses, two-, four- and five-axle trucks.

The growth limit in 2017 was found in the flow of the systems and a tendency to decrease in the use of three of the categories, except in that of three- and four-axle trucks. Simultaneous cyclic behaviors were also found in the identification and use of sinusoidal models, in which cyclicity is present.

5 Conclusion

The available information does not allow to segment the six types of database categories of the original data. Cyclical behaviors and limited growth models are observed to describe the data, so it is possible to identify the limits of the growth of this activity.

The use of transport and infrastructure services has grown in the period 2000 to 2016 with stagnation in its growth in 2016. Therefore, it is considered that some factor in the system, be it financial availability, infrastructure, or consumption, or international economic activity, limits and decreases transport consumption for several modes, aspects that can be studied in other works.

References

Francois, J., Manchin, M.: Institutions, infrastructure, and trade. World Dev. **46**, 165–175 (2013). https://doi.org/10.1016/j.worlddev.2013.02.009

Gani, A.: The logistics performance effect in international trade. Asian J. Shipping Logistics **33**(4), 279–288 (2017). https://doi.org/10.1016/j.ajsl.2017.12.012

Ghiara, H., Caminati, M.: Maritime and logistics advanced producer services within the Mediterranean: the liner shipping companies' positioning strategies in new market areas. Transp. Res. Proc. **25**, 165–186 (2017). https://doi.org/10.1016/j.trpro.2017.05.388

Grimaldi, D., Diaz, J., Arboleda, H., Fernandez, V.: Data maturity analysis and business performance. a Colombian case study. Heliyon **5**(8), e02195 (2019). https://doi.org/10.1016/j.heliyon.2019.e02195

Olariaga, O.D., Bolívar, N., Gutiérrez, R.M., Galeana, O.R.: Gravitational analysis of the air transport network. Application to the case of Colombia Transp. Res. Proc. **33**, 51–58 (2018). https://doi.org/10.1016/j.trpro.2018.10.075

Suárez-Alemán, A., Serebrisky, T., Perelman, S.: Benchmarking economic infrastructure efficiency: how does the Latin America and Caribbean region compare? Utilities Policy **58**, 1–15 (2019). https://doi.org/10.1016/j.jup.2019.03.003

Ueki, Y.: Trade costs and exportation: a comparison between enterprises in Southeast Asia and Latin America. J. Bus. Res. **68**(4), 888–893 (2015). https://doi.org/10.1016/j.jbusres.2014.11.047

Check for updates

Simultaneity Between Banking and Land Transport Activity in Colombia 2005 to 2016 Using Wavelet Techniques

Wilson A. Pinzón R.[1(✉)], Diego E. K. Rodríguez[2], and José M. Díaz P.[2]

[1] Facultad Tecnológica, Universidad Distrital "Francisco José de Caldas", Bogotá, Colombia
wapinzon@udistrital.edu.co

[2] Facultad Ciencias Empresariales, Corporación Universitaria Minuto de Dios, Bogotá, Colombia
drodr191@uniminuto.edu.co, jdiazpulido@uniminuto.edu

Abstract. This text identifies the quarterly evolution between bank activity and terrestrial transport inside Colombia since 2005 to 2016. To find causality direction their importance we use the Wavelet Graphic techniques.

Keywords: Colombia · Banks · Land transportation · Wavelet

1 Introduction

This text focuses on the identification of causality and its direction between banking activity in Colombia between 2005 and 2016 on a quarterly basis and aggregates land transport information according to the type of vehicle at the national level for 48 quarters. It searches for the relationships between the two types of variables and finds a relationship of one variable over another and a two-way causal relationship, and identifies the predominant direction.

2 Methodology

This text first reviews the background of the relationship between the physical flow of goods and the banking activity; then it takes the available public information of the credit activity and the logistic flow and finally concludes between the relationship between them based on the data available for Colombia.

3 Results

(i) Literary revision

Regarding the physical flow of goods and banking: Messaoud and Teheni (2014) seek the relationship between the relationship to business and economic growth for 162

M. T. Castañeda Galvis et al. (Eds.): ICIE 2019, LNMUINEN, pp. 180–184, 2021.
https://doi.org/10.1007/978-3-030-49370-7_20

countries in the period 2007 to 2011. For this purpose, economic growth is used as a response variable based on multiple variables that describe business regulation. They find different directions of positive, negative and neutral impact on the country's economic growth, for 162 countries during 2007 to 2011.

Guarin and Lozano (2017) develop a model to prevent banking credit crises for 4 Latin American markets based on economic activity indicators. They also develop a model for forecasting banking crises for Colombia based on the monthly banking information of the same banks between 1996 and 2013 (Lozano and Guarín 2014). On the other hand, Hamdaoui proposes a model of prediction, anticipation and selection of alternatives to bank credit crises (Hamdaoui 2016).

Boyd and colleagues propose the measurement of systematic banking crises based on economic theory and implement the effect of policies to respond to such crises (Boyd, De Nicolò and Rodionova 2019). They find that deposit insurance does not affect the probability of a systematic crisis in the banking system, but increases the probability of affecting the effectiveness of the response in selecting appropriate policies for the crisis, consistent with other studies. On the other hand, Moshirian and Wu identify different risk factors and models for developing and developed countries based on financial balance sheet information (Moshirian and Wu 2009, Tebaldi, Nguyen and Zuluaga 2018).

Regarding the factors affecting the flow of goods in South America, especially in Brazil, Botassso and his group (Bottasso et al. 2018) seek the relationship between the availability of infrastructure in seaports for international trade and its effect on both imports and exports in the 27 member territories of Brazil. In it they use the mathematical model of poisson of centers of gravity, this for the 30 most important commercial partners in the economic in the period 2009 to 2012. Its authors find as significant for the territories and their international trade the factors of: distance to maritime terminals, population of each territory, and the availability and endowment of the port, affecting even more exports than imports. They also identify an overflow effect between port infrastructure and productive activity in each territory.

Given that the mercantile movement depends on the infrastructure used, and this changes within the countries of Latin America; Alemán and his working group compare the infrastructures between multiple countries using the Data Enveloping Analysis, DEA (Suárez-Alemán, Serebrisky and Perelman 2019) and identifies different categories or groups in which they are classified with respect to 200 countries during the period 1996 to 2016.

Since logistical performance is associated with the availability and use of information technology, and in the case of Colombia, the effect of data availability on service provision

is studied, and data availability is identified as a necessary but not sufficient condition for the provision of such services, within the customer service orientation (Grimaldi et al. 2019).

(ii) available information

The detail of the movement of the bank statement for Colombia for the period 2005 to 2017, the figures indicate the level of deposits with quarterly cut in each type of account, in Table 1, in Table 2 are the figures of the composition of transport.

Table 1. Composition of elements of the Banking system in Colombia. Source, Ministry of Finance and Public Credit, Colombia. Own Calculations.

Values	2005	2011	2016
Loan portfolio	559.280.128.129.357	1.773.434.339.802.950	3.623.141.028.956.240
Savings deposit	377.299.415.482.414	868.012.583.791.254	1.506.251.017.515.840
Term deposit certificate	176.590.621.107.757	405.984.072.617.670	1.058.692.097.517.860
Deposits to current bank account	149.064.638.403.563	320.225.295.696.466	453.584.428.023.187
Home loan	70.649.122.264.184	142.311.882.068.517	477.042.081.586.408
Outstanding investment securities	19.897.213.748.362	159.908.165.781.364	337.166.890.476.498
Special savings account	2.499.739.194.618	5.357.416.192.399	10.619.022.024.595
Certificates of real value savings	13.327.739.335.950	511.550.816.473	186.496.753.777
Simple deposit	35.669.127	26.023.213.549	148.422.050.134
Total	1.368.608.653.335.330	3.675.771.329.980.640	7.466.831.484.904.540

By obtaining the correlation coefficients between the 9 banking variables and the 4 types of transport types, values greater than 0.70 are obtained in any case, identifying that each two variables are positively associated. The correlations are shown in Table 3.

Table 2. Transport composition Colombia. Source: INVIAS, Colombia. Own Calculations.

Year	Integrated System, SIPT	Subtotal buses, two-, three-, five- and six-axle	Three-axle and four-axle trucks	Cars, camperos, vans
2005	11.777.699	40.038.373	3.887.877	68.001.942
2006	14.037.417	43.592.616	4.538.910	72.128.000
2007	16.292.156	46.231.070	5.762.348	77.634.133
2008	17.678.313	46.950.029	6.959.332	79.594.166
2009	20.043.836	45.993.795	6.132.282	82.674.117
2010	23.844.309	47.355.850	7.130.984	84.471.080
2011	26.297.455	51.637.741	8.774.509	90.174.479
2012	29.403.521	48.346.206	10.564.437	115.357.355
2013	38.087.657	50.049.141	10.940.898	123.665.520
2014	48.472.803	57.153.580	12.478.758	142.893.349
2015	57.783.799	63.051.568	13.866.133	163.798.387
2016	62.621.558	62.972.469	13.495.925	170.442.521
General total	366.340.523	603.372.438	104.532.392	1.270.835.049

(iii) Cyclicality in Wavelet

Given the high correlations between the two types of banking and transport variables, then they accumulate as two totals and with these the decompositions of the sums of wavelets are made to identify the cyclicality between them, in Illustration 1.

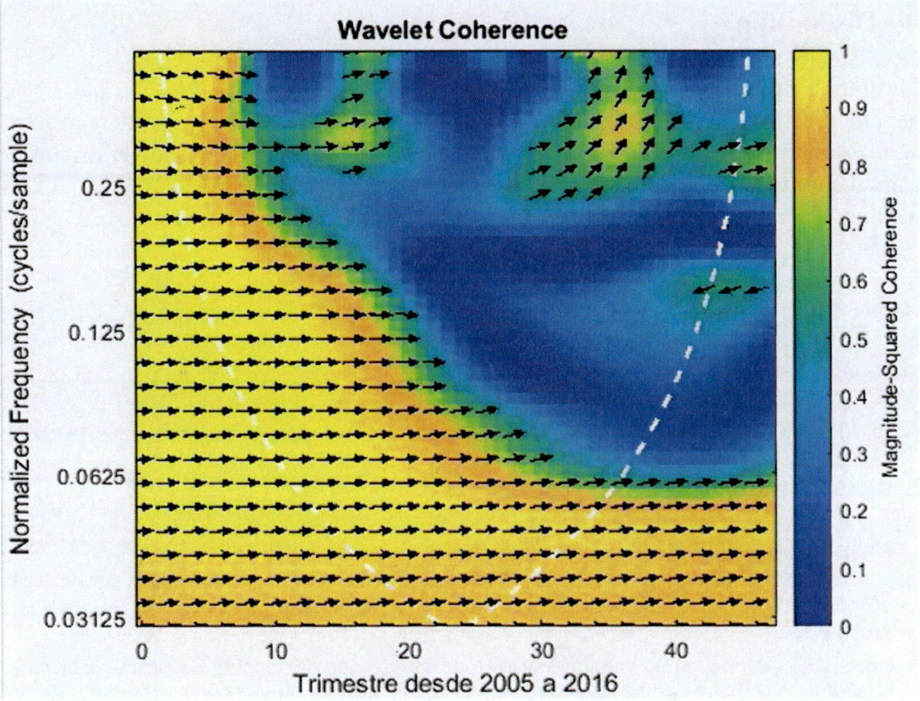

Illustration 1. Wavelet between Total transport and Total banking level. Own calculations.

Table 3. Correlation between banking and land transport variables Colombia. Own Calculations

Correlations	Integrated System, SIPT	Subtotal buses, two-, three-, five- and six- axle trucks	Three-axle and four-axle trucks	Cars, camperos, vans
Loan portfolio	0,986	0,926	0,978	0,972
Savings deposit	0,977	0,908	0,978	0,968
Term deposit certificate	0,969	0,888	0,931	0,950
Deposits to current bank account	0,940	0,892	0,968	0,939
Home loan	0,984	0,920	0,911	0,965
Outstanding investment securities	0,955	0,876	0,971	0,950
Special savings account	0,987	0,934	0,949	0,971
Certificates of real value savings	−0,663	−0,702	−0,714	−0,627
Simple deposit	0,909	0,849	0,772	0,882

4 Discussion

Illustration 1 indicates a high correlation in all frequency components and for all periods for normalized frequencies higher than 0.0625 or 1 every 16 quarters or more, up to 24 quarters, and for lower periods i.e. between 2 to 16 quarters slowly disappears l simultaneity between both types of movements the banking and the commercial flow, making it increasingly independent in the movement between the two.

References

Bottasso, A., et al.: Port infrastructures and trade: empirical evidence from Brazil. Transp. Res. Part A: Policy Pract. **107**, 126–139 (2018). https://doi.org/10.1016/j.tra.2017.11.013

Boyd, J.H., De Nicolò, G., Rodionova, T.: Banking crises and crisis dating: disentangling shocks and policy responses. J. Financ. Stab. **41**(C), 45–54 (2019). https://econpapers.repec.org/RePEc:eee:finsta:v:41:y:2019:i:c:p:45-54

Grimaldi, D., et al.: Data maturity analysis and business performance. A Colombian case study. Heliyon **5**(8), e02195 (2019). https://doi.org/10.1016/j.heliyon.2019.e02195

Guarin, A., Lozano, I.: Credit funding and banking fragility: a forecasting model for emerging economies. Emerg. Markets Rev. **32**, 168–189 (2017). https://doi.org/10.1016/j.ememar.2017.06.004

Hamdaoui, M.: Are systemic banking crises in developed and developing countries predictable? J. Multinational Financ. Manage. **37–38**, 114–138 (2016). https://doi.org/10.1016/j.mulfin.2016.09.002

Lozano, I., Guarín, A.: Banking fragility in Colombia: An empirical analysis based on balance sheets. Ensayos sobre Política Económica **32**(75), 48–63 (2014). https://doi.org/10.1016/j.espe.2014.10.001

Messaoud, B., Teheni, Z.E.G.: Business regulations and economic growth: What can be explained? Int. Strateg. Manage. Rev. **2**(2), 69–78 (2014). https://doi.org/10.1016/j.ism.2014.03.001

Moshirian, F., Wu, Q.: Banking industry volatility and banking crises. J. Int. Financ. Markets, Inst. Money **19**(2), 351–370 (2009). https://doi.org/10.1016/j.intfin.2008.02.002

Suárez-Alemán, A., Serebrisky, T., Perelman, S.: Benchmarking economic infrastructure efficiency: how does the Latin America and Caribbean region compare? Utilities Policy **58**, 1–15 (2019). https://doi.org/10.1016/j.jup.2019.03.003

Tebaldi, E., Nguyen, H., Zuluaga, J.: Determinants of emerging markets' financial health: a panel data study of sovereign bond spreads. Res. Int. Bus. Financ. **45**, 82–93 (2018). https://doi.org/10.1016/j.ribaf.2017.07.135

Author Index

M. T. Castañeda Galvis et al. (Eds.): ICIE 2019, LNMUINEN, pp. 185–186, 2021.
https://doi.org/10.1007/978-3-030-49370-7